Cracking the High School Math Competitions Solution Manual

Covering AMC 10 & 12, ARML and ZIML

Kevin Wang, Ph.D.
Kelly Ren
John Lensmire

PUBLISHED BY ARETEEM INSTITUTE

WWW.ARETEEM.ORG

ISBN: 1-944863-01-X
ISBN-13: 978-1-944863-01-2

First printing, February 2016.

Contents

1. Introduction

This book contains the solutions of the problems in the book "*Cracking the High School Math Competitions*".

The problems are included. Some minor changes are made to fix typographical errors. In a few places the problems are slightly rephrased for more clarity.

2. Number Theory

Problem 2.1 The equation

$$62 - 63 = 1$$

is obviously false. Can you move only one digit to make the resulting equation true?
<u>Solution</u>: $2^6 - 63 = 1$.

Problem 2.2 Can you find ...

(a) A multiple of 1350 that is a perfect cube? (You have *one second* to give an answer)
<u>Solution</u>: 1350^3

(b) The smallest positive multiple of 1350 that is a perfect cube?
<u>Solution</u>: $1350 = 2 \cdot 3^3 \cdot 5^2$, so the answer is $2^3 \cdot 3^3 \cdot 5^3 = 27000$.

(c) All positive multiples of 1350 that are perfect cubes?
<u>Solution</u>: $27000n^3$ for all positive integers n.

Problem 2.3 One hundred light bulbs were labeled $1,2,\ldots,100$, each controlled by one switch. At the beginning all light bulbs were turned off. One hundred students, who had nothing better to do, started flipping switches. Suppose the first student flipped every switch; the 2nd student flipped the switches of light bulbs labeled with even numbers; the 3rd student flipped the switches labeled with multiples of 3; the 4th student flipped those labeled with multiples of 4, and so on. The last student only flipped the switch

---

labeled 100. Which light bulbs are on at the end?

Solution: The perfect squares.

For any bulb, the switch is flipped only if the student number is a factor of the bulb's label. Suppose the bulb's label is n, then if a is a factor of n, so is $\frac{n}{a}$, therefore the factors of n are in pairs, unless $a = \frac{n}{a}$, in which case $n = a^2$. So the if the bulb's label n is a perfect square, its switch is flipped an odd number of times, and the bulb is on a the end; if the bulb's level is not a perfect square, its switch is slipped an even number of times, and the bulb is off at he end. Thus the answer: the bulb is on if and only if its label is a perfect square.

Problem 2.4 Diophantus was one of the last great Greek mathematicians; he developed his own algebraic notation and is sometimes called "the father of algebra." This riddle about Diophantus' age when he died was carved on his tomb:

> God vouchsafed that he should be a boy for the sixth part of his life; when a twelfth was added, his cheeks acquired a beard; He kindled for him the light of marriage after a seventh, and in the fifth year after his marriage He granted him a son. Alas! late-begotten and miserable child, when he had reached the measure of half his father's life, the chill grave took him. After consoling his grief by this science of numbers for four years, he reached the end of his life.

How long did Diophantus live? (Can you do it without algebra? Can you do it in three seconds?)

Solution: 84 years.

The paragraph describes an equation:

$$\frac{x}{6} + \frac{x}{12} + \frac{x}{7} + 5 + \frac{x}{2} + 4 = x,$$

We can solve the equation to get $x = 84$. Also, based on the information that x must be an integer that is divisible by 12 and 7, the only meaningful answer is 84 (that is the 3-second solution).

Problem 2.5 Convert:
  (a) 11001111010110_2 to base 8

    Solution: 31726.

    To convert from base 2 to base 8, we only need to split the number in 3-digit segments: $11,001,111,010,110_2$. Each segment corresponds to one digit in base 8, so the answer is 31726_8.

(b) 123_5 to base 4
Solution: 212.
Convert to base 10 first: $123_5 = 1 \times 5^2 + 2 \times 5 + 3 = 38$, then convert to base 4:
$38 = 2 \times 4^2 + 1 \times 4 + 2 = 212_4$.

(c) $ABCDEF_{16}$ to decimal
Solution: 11259375.

$$ABCDEF_{16} = 10 \times 16^5 + 11 \times 16^4 + 12 \times 16^3 + 13 \times 16^2 + 14 \times 16 + 15$$
$$= 11259375$$

(d) 276 to base 16 (or "hexadecimal")
Solution: 114.
$276 = 1 \times 16^2 + 1 \times 16 + 4$.

Problem 2.6 Find the value of base b such that the following addition is correct:

$$6651_b + 115_b = 10066_b$$

Solution: 7.
Since the digit 6 appears, the base is at least 7. The additions on the rightmost two digits are normal, but the 3rd digit from the right is $6 + 1 = 0$ with a carry. This means the base must be 7.

Problem 2.7 Find the pattern: In the sequence 110, 20, x, 11, y, z, 6, w, ... What are x, y, z, w?
Solution: 12,10,6,6.
All of these numbers are the same number 6 in different bases: $6 = 110_2 = 20_3 = 12_4 = 11_5 = 10_6 = 6_7 = 6_8 = \cdots$. Starting base 7, every term will be 6.

Problem 2.8 Explain why: $Halloween = Christmas$
Solution: 31 Oct = 25 Dec, interpreted as "31 in octal (base 8) equals 25 in decimal (base 10)". This is a well-known joke in computer science. The origin can be traced to the Isaac Asimov mystery story "The Family Man", first published in 1976.

Problem 2.9 In a store, The unit price of a certain item is an integer greater than $1, and unchanged for two years. Last year the total sale of this item was $36,963, and this year the total sale is $59,570. What is the unit price?

Solution: $37.

The unit price is a common factor of 36963 and 59570. To find the common factors of these two numbers, we find the prime factorization of both: $36963 = 3^3 \cdot 37^2$ and $59570 = 2 \cdot 5 \cdot 7 \cdot 23 \cdot 37$. The greatest common divisor of these numbers is 37. Since 37 is a prime, and the unit price is greater than 1, the only possible answer is 37.

Note: It might not be easy to find prime factors as large as 23 or 37. To find GCD of large numbers such as 59570 and 36963, it is helpful to use the Euclidean algorithm:

$$\gcd(59570, 36963) = \gcd(59570 - 36963, 36963) = \gcd(22607, 36963) = \cdots = 37.$$

Problem 2.10 Attach 3 digits after the number 503 so that the resulting 6-digit integer is a multiple of 7, 9 and 11.

Solution: 503118, and 503811.

The resulting integer should be a multiple of 693. Divide 503000 by 693, the remainder is 575, so we need to add 118 to reach the next multiple of 693. Therefore, one of the numbers is 503118. Adding another multiple of 693, we get the second solution 503811.

Problem 2.11 Find a positive integer containing all ten digits: 0,1,2,3,4,5,6,7,8,9, that is a multiple of 126.

Solution: 12345678900954. The answer is not unique, as long as it contains all the ten digits.

Problem 2.12 The integer $\overline{1a2a3a4a5a}$ is divisible by 11. What is a?

Solution: 3.

The divisibility rule for 11: $1 - a + 2 - a + 3 - a + 4 - a + 5 - a = 15 - 5a$ is a multiple of 11, so it should be 0.

Problem 2.13 A seven-digit number has seven distinct digits, and it is divisible by 11. What is the largest such number?

Solution: 9876504

Problem 2.14 A six-digit number, all of whose digits are distinct, is divisible by 11. Given that its left-most digit is 3. What is the smallest such number?

Solution: 301246

Problem 2.15 Let k be an even number. Is it possible to write 1 as the sum of the reciprocals of k odd integers?

Solution: No.

We prove this by contradiction: suppose $\dfrac{1}{n_1} + \dfrac{1}{n_2} + \cdots + \dfrac{1}{n_k} = 1$, multiply both sides by the denominators, then

$$n_2 n_3 \cdots n_k + n_1 n_3 \cdots n_k + \cdots n_1 n_2 \cdots n_{k-1} = n_1 n_2 \cdots n_k,$$

however, the left hand side is even, and the right hand side is odd, contradiction.

Problem 2.16 Find all positive integers n for which $3n - 4$, $4n - 5$, and $5n - 3$ are all prime numbers.
Solution: 2.
The number $4n - 5$ must be odd; also the parities of $3n - 4$ and $5n - 3$ are always opposite. This means that one of these numbers is even and the other two odd. There is only one even prime number, which has to be 2, so either $3n - 4 = 2$ or $5n - 3 = 2$. The former case gives $n = 2$, and the three primes numbers are 2,3,7. For the latter case, $n = 1$, and then $3n - 4 = -1$ and $4n - 5 = -1$, and those are not primes. So the only answer is $n = 2$.

Problem 2.17 (AIME 1984) The integer n is the smallest positive multiple of 15 such that every digit of n is either 0 or 8. Find n.
Solution: 8880.
It has to be a multiple of 5, so the last digit is 0. It is also a multiple of 3, so there have to be three 8s. The smallest such number is 8880.

Problem 2.18 (IMO 1959) Show that the fraction $\dfrac{12n + 1}{30n + 2}$ is irreducible for all positive integers n.
Solution: Since $5(12n + 1) - 2(30n + 2) = 1$, by Bézout's identity, $\gcd(12n + 1, 30n + 2) = 1$ no matter what n is.
Alternatively, Use Euclidean algorithm to get the GCD of the numerator and denominator:
$\gcd(12n + 1, 30n + 2) = \gcd(12n + 1, 6n) = \gcd(1, 6n) = 1$.

Problem 2.19 Determine the number of five-digit positive integers $\overline{abcde}$ (a, b, c, d, and e not necessarily distinct) such that the sum of the three-digit number $\overline{abc}$ and the two-digit number $\overline{de}$ (here d doesn't have to be nonzero) is divisible by 11.
Solution: 8181.
$\overline{abcde} = \overline{abc} \times 100 + \overline{de} = \overline{abc} \times 99 + \overline{abc} + \overline{de}$, so in fact the requirement is simply that the original five-digit number $\overline{abcde}$ is divisible by 11. The number of five-digit numbers divisible by 11 is $\lfloor 99999/11 \rfloor - \lfloor 9999/11 \rfloor = 8181$.

Problem 2.20 The number 27000001 has exactly four prime factors. Find these primes factors.

Solution: 7,43,271,331.
Use factoring formulas,

$$
\begin{aligned}
27000001 = 300^3 + 1^3 &= (300+1)(300^2 - 300 + 1) \\
&= (301)(300^2 + 2 \cdot 300 + 1 - 900) \\
&= (7 \times 43)((300+1)^2 - 30^2) \\
&= 7 \times 43(301 + 30)(301 - 30) \\
&= 7 \times 43 \times 331 \times 271.
\end{aligned}
$$

Problem 2.21 Given integers n, m, $n \geq m \geq 1$. Show that $\dfrac{\gcd(m,n)}{n}\dbinom{n}{m}$ is an integer.
(Putnam 2000)
Solution: Use Bézout's identity: There exist integers a, b so that $\gcd(m,n) = am + bn$.
Thus

$$
\begin{aligned}
\frac{\gcd(m,n)}{n}\binom{n}{m} &= \frac{am+bn}{n}\binom{n}{m} \\
&= \frac{am}{n}\binom{n}{m} + b\binom{n}{m} \\
&= \frac{am}{n} \cdot \frac{n!}{m!(n-m)!} + b\binom{n}{m} \\
&= a\frac{(n-1)!}{(m-1)!(n-m)!} + b\binom{n}{m} \\
&= a\binom{n-1}{m-1} + b\binom{n}{m}
\end{aligned}
$$

Since binomial coefficients are all integers, the original expression is an integer.

Problem 2.22 (AHSME 1976) If p and q are primes and $x^2 - px + q = 0$ has distinct positive integral roots, find p and q.
Solution: $p = 3, q = 2$.
Use Vieta's Theorem. Let x_1 and x_2 be the roots, $x_1 x_2 = q$ is a prime, so one of the roots is 1 and the other is not 1. Assume $x_1 = 1$ and then $x_2 = q$. We get that $p = x_1 + x_2 = q + 1$. The only two primes that are consecutive numbers are: $p = 3, q = 2$.

Problem 2.23 Given a six-digit number $\overline{abcdef}$, whose digits are 1,2,3,4,5,6, not necessarily in this order. Assume that $6 \mid \overline{abcdef}$, $5 \mid \overline{abcde}$, $4 \mid \overline{abcd}$, $3 \mid \overline{abc}$, and $2 \mid \overline{ab}$. Find $\overline{abcdef}$.
Solution: 123654, 321654.
It can be determined immediately that $e = 5$.
Also we know that b, d, f must be even digits, and thus a, c are odd. So either $a = 1, c = 3$ or $c = 1, a = 3$.

To make $\overline{abc}$ a multiple of 3, only $b=2$ works.

To determine d, note that we need to have $4 \mid \overline{cd}$, and it could be 16 or 36, but not 14 or 34.

So $d=6$ and then $f=4$. Hence there are two possible solutions.

Problem 2.24 Compute the product of all distinct positive divisors of 120^6 (express your answer as a power of 120).

Solution: 120^{2793}.

Note that factors come in pairs. For any number n, if a is a factor, n/a is also a factor, and the product of these two factors is n. The only possible non-paired factor is the square root of a perfect square. For 120^6, it is a perfect square. The prime factorization $120^6 = 2^{18}3^6 5^6$, so there are $(18+1)(6+1)(6+1) = 931$ factors, with 465 pairs and one square root. The product of all these factors is $(120^6)^{465} 120^3 = 120^{2793}$.

Problem 2.25 Verify the following facts: Let n be an integer, then:

(a) $n^2 \equiv 0$ or $1 \pmod 3$;

(b) $n^2 \equiv 0$ or $\pm 1 \pmod 5$;

(c) $n^2 \equiv 0$ or 1 or $4 \pmod 8$;

(d) $n^3 \equiv 0$ or $\pm 1 \pmod 9$;

(e) $n^4 \equiv 0$ or $1 \pmod{16}$;

Solution: List all possible remainders k for each modulus, and calculate the remainders of k^2 or k^3 and find out the possible results.

Problem 2.26 For perfect squares, not all values in a certain moduli are possible remainders.

(a) Find the possible remainders of n^2 in $\pmod 4$.
   Solution: 0,1

(b) Find the possible remainders of n^2 in $\pmod 9$.
   Solution: 0,1,4,7

Problem 2.27 If $m > 1$ and $69 \equiv 90 \equiv 125 \pmod m$, what is m?

Solution: $90 - 69 = 21, 125 - 90 = 35$, so $m \mid \gcd(21,35) = 7$, thus $m = 7$.

Problem 2.28 What is the remainder of $2^{50} + 3^{50}$ when divided by 13?

Solution: 0.

Apply the Fermat's Little Theorem, $2^{12} \equiv 1 \pmod{13}$, so $2^{50} \equiv (2^{12})^4 2^2 \equiv 4 \pmod{13}$. Similarly, $3^{50} \equiv 9 \pmod{13}$. Adding 4 and 9, we get $0 \pmod{13}$.

Second Solution: Using factoring formula, $2^{50} + 3^{50} = 4^{25} + 9^{25} = (4+9)(\cdots) = (13)(\cdots)$ is a multiple of 13. So the remainder is 0.

Problem 2.29 Show that there are no perfect squares in the sequence:

$$11, 111, 1111, 11111, \ldots$$

Solution: All of them are 3 (mod 4), but squares are all 0 or 1 mod 4.

Problem 2.30 Is it possible to find two integers n and m such that $n^2 + m^2 = 2015$?
Solution: No.
The remainder of any square divided by 4 is either 0 or 1. Thus the sum of two squares $n^2 + m^2$ can be 0, 1, or 2 in mod 4. However, $2015 \equiv 3 \pmod 4$, thus is it not possible.

Problem 2.31 Can a 5-digit number consisting only of distinct even digits be a perfect square?
Solution: No.
The number consists of the digits $0, 2, 4, 6, 8$, one each. In mod 9, the squares have remainders 0, 1, 4, or 7. This number has remainder 2, no matter how the digits are arranged, so it is not a square.

Problem 2.32 The number 2^{29} is a nine-digit number all of whose digits are distinct. Without computing the actual number, determine which of the ten digits is missing.
Solution: 4.
$2^{29} \equiv 5 \pmod 9$ using patterns or Fermat's Little Theorem. If all digits were present, the sum of digits would have been $0 + 1 + \cdots + 9 = 45$, a multiple of 9. One digit is missing and we get 5 mod 9, so it is 4 that's missing.

Problem 2.33 Show that in the set of 7! numbers consisting of the distinct permutations of the digits $1, 2, 3, 4, 5, 6, 7$, no member is a multiple of another.
Solution: Consider the numbers in mod 9. All of them are 1 mod 9. If one is a multiple of another, it would be 2 or 3 or 4 or 5 or 6 or 7 times the other, not possible to remain 1 mod 9.

Problem 2.34 A certain natural number n has a unit digit 9 when expressed in base 12. Find the remainder when n^2 is divided by 6.
Solution: 3.
$n = 12k + 9$, so $n^2 = (12k+9)^2 = 144k^2 + 216k + 81$. If n^2 is divided by 6, the remainder is 3.

Problem 2.35 Find the remainder when

$$37 + 377 + 3777 + 37777 + \cdots + 3777777777777777$$

is divided by 11 (note the last summand is 16 digits long (one 3 and fifteen 7's).
Solution: 9.
$37 \equiv 4 \pmod{11}$, $377 \equiv 3 \pmod{11}$, etc. The pattern of remainders of each term divided by 11 is: $4 + 3 + 4 + 3 + \cdots + 4$. There are eight 4s and seven 3s, so the sum is 53, which is 9 mod 11.

Problem 2.36 The Fibonacci sequence is defined by $F_1 = F_2 = 1$, and $F_{n+2} = F_{n+1} + F_n$, that is, the first two terms are both 1, and each subsequence term is the sum of the previous two terms. Find the remainder when F_{2011} is divided by 7.
Solution: 5.
Calculate the terms in mod 7, the pattern repeats every 16 terms:

$$1, 1, 2, 3, 5, 1, 6, 0, 6, 6, 5, 4, 2, 6, 1, 0, 1, 1, \ldots.$$

$2011 \equiv 11 \pmod{16}$ so we look at the 11th term, which is 5.

Problem 2.37 How many zeros are there at the end of 1000! ?
Solution: 249.
Each 0 at the end is made by a factor 2 and a factor 5. In 100! there are more factors of 2 than 5, so we just need to count the factors 5. There are 200 multiples of 5, each contributing one 0; there are also 40 multiples of 25, each contributing one additional 0; also, 8 multiples of 125, each contributing one additional 0; finally there is one multiple of 625, contributing yet another 0. Totally, there are $200 + 40 + 8 + 1 = 249$ zeros.

Problem 2.38 There are two two-digit numbers whose square ends in the same two-digit number. Find them.
Solution: 25 and 76.
The ones digit has to be a 5 or a 6. If the number is $10x + 5$, then $(10x + 5)^2 = 100x^2 + 100x + 25$, so the last two digits are 25. If the number is $10x + 6$, then $(10x + 6)^2 = 100x^2 + 120x + 36 = 100(x^2 + x) + 20x + 30 + 6$. Now since $20x + 30 \equiv 10x \pmod{100}$, we get $10x + 30 \equiv 0 \pmod{100}$, which means $x = 7$, so 76 is the other two-digit number that ends in 76 when squared.

Problem 2.39 Find all Pythagorean triples containing the number 29.
Solution: (20,21,29); (29,420,421)
As we know, all primitive Pythagorean triples (a, b, c) can be obtained from the formula: $a = m^2 - n^2, b = 2mn, c = m^2 + n^2$ where m, n are integers such that $\gcd(m, n) = 1$.

The number 29 can be a or c. It cannot be b since b is even. If $a = m^2 - n^2 = 29$, then $(m+n)(m-n) = 29$. Since 29 is prime, we have $m+n = 29$ and $m-n = 1$. Thus $m = 15, n = 14$, and we get $a = 29, b = 420, c = 421$. If $c = m^2 + n^2 = 29$, the only possibility is $m = 5, n = 2$. Thus $a = 21, b = 20, c = 29$.

Problem 2.40 Find all Pythagorean triples containing the number 15. (Note: there are 5 such triples.)

Solution: $(8,15,17)$; $(15,112,113)$; $(15,20,25)$; $(9,12,15)$; $(15,36,39)$.

As we know, all primitive Pythagorean triples (a,b,c) can be obtained from the formula: $a = m^2 - n^2, b = 2mn, c = m^2 + n^2$ where m,n are integers such that $\gcd(m,n) = 1$.

The number 15 is not prime, so the Pythagorean triple could either be a primitive one containing 15, or a multiple of a primitive one containing 3, or a multiple of a primitive one containing 5.

Case 1. The Pythagorean triple is primitive containing 15. Here 15 cannot be c, otherwise we would have $m^2 + n^2 = 15$, but since squares can only be 0 or 1 in mod 4, the sum $m^2 + n^2$ can be 0 or 1 or 2, but not 3 mod 4, and yet $15 \equiv 3 \pmod 4$. Thus $a = 15$, so $15 = m^2 - n^2 = (m+n)(m-n)$. There are two possibilities: $m+n = 15, m-n = 1$, thus $m = 8, n = 7$ and the triple is $(15, 112, 113)$; or $m+n = 5, m-n = 3$, thus $m = 4, n = 1$ and the triple is $(15, 8, 17)$.

Case 2. The Pythagorean triple is a multiple of a primitive Pythagorean triple containing 3. Again, 3 can only be a. So $m^2 - n^2 = 3$, then $m+n = 3, m-n = 1$, and $m = 2, n = 1$, so the primitive triple is $(3,4,5)$, and the triple containing 15 is $(15,20,25)$.

Case 3. The Pythagorean triple is a multiple of a primitive Pythagorean triple containing 5. Now 5 can be a or c. If $a = m^2 - n^2 = 37$, then $(m+n)(m-n) = 5$. Since 5 is prime, we have $m+n = 5$ and $m-n = 1$. Thus $m = 3, n = 2$, and we get $a = 5, b = 12, c = 13$, so the triple containing 15 is $(15,36,39)$. If $c = m^2 + n^2 = 5$, the only possibility is $m = 2, n = 1$. Thus $a = 3, b = 4, c = 5$, so the triple containing 15 is $(9,12,15)$.

Problem 2.41 Are there any prime numbers between 2020 and 2030?

Solution: Yes, 2027 and 2029 are prime.

$2020 = 2 \times 5 \times 101$
$2021 = 43 \times 47$
$2022 = 2 \times 3 \times 337$
$2023 = 7 \times 17^2$
$2024 = 2^3 \times 11 \times 23$
$2025 = 3^4 \times 5^2$
$2026 = 2 \times 1013$
$2028 = 2^2 \times 3 \times 13^2$
$2030 = 2 \times 5 \times 7 \times 29$

Problem 2.42 An 8-digit number $\overline{141A28B3}$ is a multiple of 99. Find A and B.
Solution: $A = B = 4$. Use the rules of divisibility for 9 and 11. Since the number is divisible by 9, the sum of the digits is divisible by 9. Thus $1+4+1+A+2+8+B+3 = 19+A+B \equiv 1+A+B \equiv 0 \pmod 9$. Thus $A+B = 8$ or 17. By divisibility by 11, the alternating sum of the digits is divisible by 11, thus $1-4+1-A+2-8+B-3 = B-A-11 \equiv B-A \equiv 0 \pmod{11}$, and the only possibility is $A = B$. So the final conclusion is $A = B = 4$.

Problem 2.43 Find the smallest six digit integer with leading digit 7, and all digits are distinct, that is divisible by 11.
Solution: 701239.
We start with the smallest possible leading digits: assume the number to be $70123x$, where x is a digit. Using the divisibility rule of 11, $7-0+1-2+3-x = 9-x \equiv 0 \pmod{11}$. This means $x = 9$.
Note: in the case where no digit could work as x, we would modify the assumption to work with $70124x$ and find an x to satisfy the requirement.

Problem 2.44 Find the smallest positive integer satisfying both of the following requirements:
  (a) Its units digit is 6;
  (b) If the units digit 6 is moved before the first digit, the new number is 4 times the original number.

Solution: 153846.
Just work out the digits one by one, starting from the right.

Problem 2.45 Find the smallest positive multiple of 225, all of whose digits are 0 or 1 in its base 10 representation.
Solution: 11111111100.
This number requires two 0s at the end to ensure that the number is a multiple of 25, and there have to be 9 ones to make the number a multiple of 9. The smallest is 11111111100.

Problem 2.46 Find the smallest positive integer n such that $\sqrt{2000n}$ is an integer.
Solution: 5.
Since $2000 = 2^4 5^3$, we need one more 5 to make the product a square.

Problem 2.47 Given that $2^{96} - 1$ is divisible by two integers between 60 and 70. What are these two integer?
Solution: 63,65.

Factor using difference of squares, until getting $(2^6 + 1)(2^6 - 1)$.

Problem 2.48 Find the greatest common divisor of the following ten integers: $2000^3 + 3 \cdot 2000^2 + 2 \cdot 2000$, $2001^3 + 3 \cdot 2001^2 + 2 \cdot 2001$, ..., $2008^3 + 3 \cdot 2008^2 + 2 \cdot 2008$, $2009^3 + 3 \cdot 2009^2 + 2 \cdot 2009$.
Solution: 6.
The expression $n^3 + 3n^2 + 2n = n(n+1)(n+2)$, and the product of any 3 consecutive integers is always a multiple of 6. To see that there are no larger common divisors for these 10 numbers: 4 or 9 is not a factor of $2001 \times 2002 \times 2003$, and no other prime factors appear in all these 10 numbers.

Problem 2.49 The six-digit number $\overline{xy342z}$ is divisible by 396. Find all such numbers.
Solution: 453420, 413424, 373428.
$396 = 4 \times 9 \times 11$, so we check divisibility by 4, 9, and 11. Divisibility by 4 means the last two digits must be divisible by 4, and there are three cases for z: 0, 4, or 8. For each of these cases, x and y are determined by rules of divisibility by 9 and 11.

Problem 2.50 Find the largest multiple of 11 among the nine-digit numbers, whose digits are all distinct.
Solution: 987652413.
We try to put the largest digits in the front, and use trial and error, based on the divisibility rule for 11.

Problem 2.51 Find all numbers n less than 50 with the following property: the product of the divisors of n is equal to n^2.
Solution: The factors come in pairs: if a is a factor of n, so is $\dfrac{n}{a}$. Each pair has a product of n itself. So, the factors 1 and n is one pair, and there would be another pair if n is not 1. That means exactly 4 factors. Therefore, n is either 1, or the product of 2 primes (pq), or the cube of a prime (p^3). So the possible numbers are: 1,6,8,10,14,15,21,22,26,27,33,34,35,38,39,46.

Problem 2.52 (1984 ARML I-7) Find all positive integers n less than twenty such that $49 \mid n! + (n+1)! + (n+2)!$.
Solution: 5,12,14,15,16,17,18,19.
$n! + (n+1)! + (n+2)! = n!(n+2)^2$, and the answer n should satisfy: either $n+2$ is a multiple of 7, or $n!$ contains two factors of 7 (which means $n \geq 14$).

Problem 2.53 What is the largest natural number k such that $\dfrac{1001 \cdot 1002 \cdots 2000}{11^k}$ is an integer?

Solution: 100.

There are 91 multiples of 11 (from $1001 = 11 \times 91$ to $1991 = 11 \times 181$), each contributing one 11; 8 multiples of 121 ($1089 = 121 \times 9$ to $1936 = 121 \times 16$), each contributing one additional 11; and 1 multiple of 1331, contributing yet another additional 11.

Problem 2.54 What is the remainder when 9^{2012} is divided by 11?

Solution: 4.

This can be solved by using patterns of powers of 9 in mod 11: the cycle length is 10. Or, use Fermat's Little Theorem, $9^{10} \equiv 1 \pmod{11}$, thus $9^{2012} \equiv 9^2 \equiv 81 \equiv 4 \pmod{11}$.

Problem 2.55 What are the last two digits of 2012^{2012}?

Solution: 56.

The last two digits of 2012^{2012} are the same as the last two digits of 12^{2012}. To avoid the big multiplications, just compute the last two digits of the powers of 12, and the pattern repeats after 20 terms:

Powers of 12	Last two digits
12^1	12
12^2	44
12^3	28
12^4	36
12^5	32
12^6	84
12^7	08
12^8	96
12^9	52
12^{10}	24
12^{11}	88
12^{12}	56
12^{13}	72
12^{14}	64
12^{15}	68
12^{16}	16
12^{17}	92
12^{18}	04
12^{19}	48
12^{20}	76
12^{21}	12

Therefore the result is the 12th entry in the pattern: 56.

Problem 2.56 A number m is the smallest positive integer that gives remainder 1 when divided by 3, remainder 5 when divided by 7, and remainder 4 when divided by 11. What is the remainder when m is divided by 4?
Solution: 3.
Try to satisfy the remainders for 3 and 7 first: the number plus 2 is a common multiple of 3 and 7, so the number is $21k - 2$. Then try $k = 1, 2, \ldots$ so that $21k - 6$ is a multiple of 11, which works when $k = 5$, and $m = 103$. and the remainder is 3.

Problem 2.57 The Fibonacci sequence is defined by $F_1 = F_2 = 1$, and $F_{n+2} = F_{n+1} + F_n$, that is, the first two terms are both 1, and each subsequent term is the sum of the previous two terms. Find the remainder when F_{2010} is divided by 7.
Solution: 6.
The pattern of remainders of the Fibonacci numbers divided by 7 is

$$1, 1, 2, 3, 5, 1, 6, 0, 6, 6, 5, 4, 2, 6, 1, 0, 1, 1, \ldots.$$

where each cycle contains 16 numbers. Since $2010 = 16 \times 125 + 10$, the answer is the 10th entry in the pattern, which is 6.

Problem 2.58 A four digit number minus the sum of its digits, the result is $\overline{20d0}$. What is d?
Solution: 7.
Any number and the sum of its digits are congruent mod 9. Thus their difference, $\overline{20d0}$, must be a multiple of 9, so $2 + d$ is a multiple of 9, and $d = 7$ is the only choice.

Problem 2.59 A five digit number $\overline{4a77b}$ is divisible by 99, find the values of a, b.
Solution: $a = 1, b = 8$.
Use the divisibility rules for 9 and 11.

Problem 2.60 A two digit number equals the sum of its tens digit and the square of its units digit. What is this two digit number?
Solution: 89.
Set up equation $10a + b = a + b^2$, so $9a = b(b-1)$. Thus either $9 \mid b$ or $9 \mid (b-1)$, which means $b = 9, 0$, or 1. Only $b = 9$ works, and $a = 8$.

Problem 2.61 (1980 Canada) Let $\overline{a679b}$ be a five-digit number. If $72 \mid \overline{a679b}$, find the values of a and b.
Solution: $a = 3, b = 2$.
It is a multiple of 8, looking at the three digits $\overline{79b}$, so $b = 2$. The sum of digits is a multiple of 9, so $a + 6 + 7 + 9 + 2 = a + 24$ is a multiple of 9. The only digit that works is $a = 3$.

Problem 2.62 Let n be a positive integer, such that $n+3$ is a multiple of 5, and $n-3$ is a multiple of 6. Find the smallest such n.
Solution: 27.
In fact $n+3$ is a common multiple of 5 and 6. The smallest positive common multiple of 5 and 6 is 30, therefore n is 27.

Problem 2.63 Factorial with trailing zeros.

(a) (1985 NYSML T-3) For how many positive integral values of n does $n!$ end with precisely 25 zeros? What are they?
Solution: 5. They are: 105,106,107,108,109.
Note that each zero means a pair of factors of 5 and 2. In $n!$, the factor 2 appears more frequently than 5, so we only need to count the factor 5. We make an educated guess of the number n: $n = 100$. The number of zeros for 100! is calculated as follows. There are 20 multiples of 5 from 1 to 100, each contributing one zero. However, we didn't take into account the multiples of 25, each of which actually contributes two zeros, but we only counted one. So we have to add one zero for each multiple of 25: there are 4 of them. Therefore 100! has 24 zeros. The problem asks for 25 zeros, so we need one more factor 5, which occurs in 105. So the values 105, 106, 107, 108, 109 all give 25 zeros at the end. The number 110 gives another 0, so 110 and above are out. Final answer: 5 numbers, 105 through 109.

(b) Same question, but what if $n!$ is represented in base eight?
Solution: None.
We count the number of factor 2s. Three 2s makes a zero in base 8. An educated guess brings us to the seventies. We will guess 79. 79! has 39 multiples of 2, 19 multiples of 4, 9 multiples of 8, 4 multiples of 16, 2 multiples of 32, and 1 multiple of 64, thus the maximum power of 2 is $2^{39+19+9+4+2+1} = 2^{74}$, and that account for 24 zeros. The next factorial, 80!, adds 4 more 2s to get 2^{78}, which means 26 zeros. There are no factorials with 25 zeros in base 8.

Problem 2.64 Find the remainder when 1996^{2000} is divided by 29.
Solution: 7.
First, $1996 \equiv 24 \equiv -5 \pmod{29}$. So $1996^{2000} \equiv (-5)^{2000} \equiv 5^{2000} \pmod{29}$. Based on Fermat's Little Theorem, $5^{28} \equiv 1 \pmod{29}$, therefore $5^{2000} \equiv (5^{28})^{71} \cdot 5^{12} \equiv 5^{12} \pmod{29}$. Then $5^{12} \equiv 25^6 \equiv 625^3 \equiv 16^3 \equiv 7 \pmod{29}$. The key is to use mod 29 to reduce the calculations whenever possible.

Problem 2.65 Find the remainder when 2001^{2000} is divided by 49.

Solution: 36.

Since 49 is not a prime, the Fermat's Little Theorem does not apply. First $2001 \equiv 41 \equiv -8 \pmod{49}$, thus $2001^{2000} \equiv (-8)^{2000} \equiv 8^{2000} \pmod{49}$. Calculate the patterns of powers of 8 in mod 49: 8, 15, 22, 29, 36, 43, 1, repeating every 7 terms, and $2000 \equiv 5 \pmod 7$, therefore $2001^{2000} \equiv 36 \pmod{49}$.

Problem 2.66 Find the remainder when $2222^{5555} + 5555^{2222}$ is divided by 7.
Solution: 0.

Fermat's Little Theorem says that $a^6 \equiv 1 \pmod 7$ if $7 \nmid a$. So, $2222^{5555} \equiv 3^{5555} \equiv 3^{6 \times 925 + 5} \equiv 3^5 \equiv 5 \pmod 7$, and $5555^{2222} \equiv 4^{2222} \equiv 4^{6 \times 370 + 2} \equiv 4^2 \equiv 2 \pmod 7$, so the sum is 0 (mod 7).

Problem 2.67 Let x, y be positive integers, $x < y$, and $x + y = 667$. Given that $\dfrac{\text{lcm}(x,y)}{\gcd(x,y)} = 120$. Find all such pairs (x, y).
Solution: $(115, 552), (232, 435)$.

Let $c = \gcd(x, y)$, and $x = ac, y = bc$, where $\gcd(a, b) = 1$, and $a < b$. Clearly $\text{lcm}(x, y) = abc$. From the given, $ab = 120$. Also, $x + y = (a + b)c = 667 = 23 \times 29$. There are four cases:

(1) $a + b = 1, c = 667$; this is not possible because a and b should be positive integers.
(2) $a + b = 667, c = 1$; this does not give integer roots for a and b.
(3) $a + b = 23, c = 29$; combined with $ab = 120$ and $a < b$, we have $a = 8, b = 15$, so $x = 232, y = 435$.
(4) $a + b = 29, c = 23$; combined with $ab = 120$ and $a < b$, we have $a = 5, b = 24$, so $x = 115, y = 552$.

Problem 2.68 Find all ordered pairs of positive integers (x, y) such that $1! + 2! + 3! + \cdots + x! = y^2$.
Solution: $(1, 1), (3, 3)$.

It is easy to see that $(1, 1)$ and $(3, 3)$ are solutions, and $x = 2$ does not give a solution. Based on the fact that $n!$ is a multiple of 10 if $n \geq 5$, and $1! + 2! + 3! + 4!$ has a last digit 3, we conclude: For $x \geq 4$, the last digit of $1! + 2! + 3! + \cdots + x!$ is always 3. The last digit of a perfect square can be 0,1,4,5,6,9, so there are no more pairs (x, y) that satisfy the equation.

Problem 2.69 In the Cartesian coordinate system, how many grid points (x, y) satisfy $(|x| - 1)^2 + (|y| - 1)^2 < 2$?
Solution: 16.

Let $a = |x| - 1$ and $b = |y| - 1$. So $a^2 + b^2 < 2$. As a, b are integers, $|a| \leq 1$ and $|b| \leq 1$, where equalities cannot hold simultaneously. We discuss the cases separately.

---

(1) $a = -1$. In this case x must be 0, and $b = 0$, so $y = \pm 1$, there are two lattice points $(0, -1)$ and $(0, 1)$.

(2) $a = 0$, so $x = \pm 1$. In this case, b can be $-1, 0, 1$. For $b = -1$, y must be 0, giving two lattice points $(-1, 0)$ and $(1, 0)$. For $b = 0$, $y = \pm 1$, so there are four points: $(\pm 1, \pm 1)$. For $b = 1$, $y = \pm 2$, also giving four lattice points: $(\pm 1, \pm 2)$.

(3) $a = 1$, in this case $b = 0$, and so $x = \pm 2$, $y = \pm 1$, giving four lattice points $(\pm 2, \pm 1)$. Putting all together, there are 16 lattice points.

Problem 2.70 Find all possible positive integers n such that $323 \mid 20^n + 16^n - 3^n - 1$.
Solution: All positive even numbers.
Since $323 = 17 \times 19$, we test for 17 and 19 separately. First, $20^n - 3^n$ is a multiple of 17 for all n. And $16^n - 1 \equiv (-1)^n - 1 \pmod{17}$, so $16^n - 1$ is a multiple of 17 if and only if n is even. Therefore, $20^n + 16^n - 3^n - 1$ is divisible by 17 if and only if n is even. So now we only consider even numbers n for divisibility by 19. $20^n - 1$ is always a multiple of 19; and $16^n - 3^n \equiv (-3)^n - 3^n \equiv 3^n - 3^n \equiv 0 \pmod{19}$ for all even numbers n. Therefore, all positive even numbers n satisfy the requirement.

Problem 2.71 Let p and d be positive integers, and $6 \nmid d$. Assume that $p, p+d, p+2d$ are all prime numbers. Then $p + 3d$ must be ... (select all that apply)
(A) prime   (B) multiple of 9   (C) multiple of 3   (D) either prime or multiple of 9.
Solution: (C).
None of $p, p+d, p+2d$ can be 2, so they are all odd prime numbers, which means d is even. Since d is not a multiple of 6, in particular, it is not a multiple of 3, therefore $p, p+d, p+2d$ must cover all possible remainders when divided by 3. That means, one of them must be a multiple of 3. However, the only way for a prime to be a multiple of 3 is that it is equal to 3 itself. Only p can be equal to 3. So we can see that $p + 3d = 3(d+1)$ must be a multiple of 3. Must it be a multiple of 9 in all cases? One example of the three primes are $3, 7, 11$ where $d = 4$. In this case $p + 3d = 15$, which is a multiple of 3 but not 9. Therefore (C) is the only answer.

Problem 2.72 Find all possible positive integers n such that $2^n - 1$ is a multiple of 7.
Solution: $3 \mid n$.
Consider n in mod 3. If $n = 3k$, then $2^{3k} - 1 = 8^k - 1 \equiv 0 \pmod 7$, so $n = 3k$ are solutions. If $n = 3k+1$, $2^{3k+1} - 1 \equiv 8^k \cdot 2 - 1 \equiv 1 \cdot 2 - 1 = 1 \pmod 7$, not a multiple of 7. If $n = 3k+2$, $2^{3k+2} - 1 \equiv 4 - 1 \equiv 3 \pmod 7$, also not a multiple of 7. So $n = 3k$ are the only solutions.

Problem 2.73 Let A be the sum of the digits of 5^{10000}, and B be the sum of digits of A, and C be the sum of digits of B. What is C?

Solution: 4.

The sum of the digits of a positive integer has the same remainder as the original integer when divided by 9. So $5^{10000} \equiv A \equiv B \equiv C \pmod 9$. Using patterns, we find that $5^{10000} \equiv 4 \pmod 9$. Now we estimate A, B, and C. Since $5^{10000} < 10^{10000}$, it has at most 10000 digits, thus $A < 9 \times 10000 = 90000$. The maximum possible sum of digits for numbers under 90000 occurs at the number 89999, therefore $B < 8+9+9+9+9 = 44$. Below 44, the maximum sum of digits occurs at 39 with a sum 12, so $C < 12$. There is only one value between 1 and 12 that is 4 (mod 9), which is 4. Hence $C = 4$.

Problem 2.74 Given a sequence: 1,4,8,10,16,19,21,25,30,43. If a group of consecutive terms in this sequence has a sum that is a multiple of 11, then call this group a "fine" group. How many "fine" groups are there?

Solution: 7.

Write the sequence in mod 11: $1, 4, -3, -1, 5, -3, -1, 3, -3, -1$. Calculate the partial sums:

$$
\begin{aligned}
S_1 &= 1 \\
S_2 &= 1+4 = 5 \\
S_3 &= 1+4+(-3) = 2 \\
S_4 &= 1+4+(-3)+(-1) = 1 \\
S_5 &= 1+4+(-3)+(-1)+5 = 6 \\
S_6 &= 1+4+(-3)+(-1)+5+(-3) = 3 \\
S_7 &= 1+4+(-3)+(-1)+5+(-3)+(-1) = 2 \\
S_8 &= 1+4+(-3)+(-1)+5+(-3)+(-1)+3 = 5 \\
S_9 &= 1+4+(-3)+(-1)+5+(-3)+(-1)+3+(-3) = 2 \\
S_{10} &= 1+4+(-3)+(-1)+5+(-3)+(-1)+3+(-3)+(-1) = 1
\end{aligned}
$$

The fine groups are the ones formed by the differences of two equal values in the partial sums. For example, $S_1 = S_4$, so the 2nd through 4th numbers form a fine group: $\{4, 8, 10\}$. Another example: $S_2 = S_8$, therefore the 3rd through 8th numbers form a fine group: $\{8, 10, 16, 19, 21, 25\}$

The values $S_1 = S_4 = S_{10} = 1$, and they form 3 fine groups. $S_2 = S_8 = 5$, and they form 1 fine group. Also, $S_3 = S_7 = S_9 = 2$, and they form 3 fine groups. So totally there are 7 fine groups.

Problem 2.75 In $\triangle ABC$, all three sides have integer lengths. Assume that $AB = 21$, and its perimeter is 54. Also known that its area is a positive integer. What are BC and CA?

Solution: 13 and 20.

Without loss of generality, assume $BC \le CA$, let $x = BC$, so $x \le 16$. By Heron's formula, the area $[ABC] = \sqrt{27 \cdot (x-6) \cdot 6 \cdot (27-x)} = 9\sqrt{2(x-6)(27-x)}$ which should be a positive integer, thus $7 \leq x \leq 16$. Checking for each value, $x = 13$ works out. Then the remaining side is 20.

Problem 2.76 Given a positive integer n, how many positive integers a are there such that $n^6 + 3a$ is a perfect cube? Justify your answer.
Solution: Infinitely many.
Consider $(n^2 + 3k)^3$ for arbitrary k: $(n^2 + 3k)^3 = n^6 + 9kn^4 + 27k^2n^2 + 27k^3$, so let $a = 3kn^4 + 9k^2n^2 + 9k^3$, then $n^6 + 3a$ is a perfect cube for all k.

Problem 2.77 From the set $\{1, 2, \ldots, 100\}$, select k numbers. What is the minimum value of k such that it is guaranteed to have two numbers that are not relatively prime?
Solution: 27.
There are 25 primes under 100. Divide the set $\{1, 2, \ldots, 100\}$ into 26 categories: the number 1 itself is a category, and each of the other categories contain the multiples of one of the primes. It is noted that these categories overlap, but together they cover all the set $\{1, 2, \ldots, 100\}$. If we choose 27 numbers, it is guaranteed that there must be two of the numbers that belong to the same category, so these two numbers are not relatively prime. So 27 is sufficient. However, if we only select 26 numbers, the worst case scenario is that we have chosen all the prime numbers plus the number 1, and all those are pairwise relatively prime. Therefore 27 is the minimum value.

Problem 2.78 For positive integer k, let $M = 2(2k - 1)$, which of the following must be true?
  (a) M is not a perfect square for any k.
  (b) There are infinitely many k such that M is a perfect square.
  (c) There is a unique k such that M is a perfect square.
  (d) There are finitely many, but more than 1, values of k such that M is a perfect square.

Solution: The first statement is always true.
Since $2k - 1$ is an odd number, M is a multiple of 2 but not a multiple of 4, therefore M cannot be a perfect square. The first statement is always true. As a consequence, none of the other statements is true.

Problem 2.79 Put a positive or negative sign in front of each of $1, 2, 3, \ldots, 2003$, and take the sum, then this sum must be
(A) Odd   (B) Even   (C) multiple of 3   (D) none of those
Solution: (B).
Changing the sign in front of any of the numbers will change the sum by an even number amount, so the parity of the sum is always the same. On the other hand it cannot always

be a multiple of 3 by considering using a plus or minus in front of 1. So now we need to determine if it is always even or always odd by assuming all signs are pluses. The sum $1 + 2 + \cdots + 2003 = (2004 \times 2003)/2$ is even, so the sums with different signs are always even.

Problem 2.80 Given that a, b, n are positive integers. Assume that for any positive integer $k \neq b$, $(k-b) \mid (k^n - a)$, then which of the following must be true?
(A) $a > b^n$ (B) $a < b^n$ (C) $a = b^n$ (D) It depends.
Solution: (C).
Clearly, $(k-b) \mid (k^n - b^n)$. Assume $a \neq b^n$, then we can find a k such that $(k-b) \nmid (a - b^n)$, but then $(k-b) \nmid (k^n - a)$, a contradiction. Therefore, $a = b^n$.

Problem 2.81 Find all ordered triples (x, y, z) of prime numbers satisfying equation $x(x+y) = z + 120$.
Solution: $(2, 59, 2)$, and $(11, 2, 23)$.
Consider the following cases for prime number z:
(1) If $z = 2$, $x(x+y) = 122 = 2 \times 61$. Thus $x = 2, y = 59$.
(2) If z is an odd prime, then $x(x+y)$ is an odd number. That means both x and $x+y$ are odd, and so $y = 2$. The equation becomes $x(x+2) = z + 120$, and so $x^2 + x - 120 = z$, so $(x+12)(x-10) = z$. But z is prime, then $x - 10 = 1$, so $x = 11, z = 23$.

Problem 2.82 Three distinct positive integers a, b, c are pairwise relatively prime, and the sum of any two is a multiple of the third one. What is the product abc?
Solution: 6.
Without loss of generality, assume $a > b > c$. So $2a > b > c = ar$, thus $r = 1$. Therefore $a = b + c$. Also, $a + c = qb$, so $b + 2c = qb$, and then $2c = (q-1)b < 2b$, Therefore $q = 2$, and $b = 2c, a = 3c$. Also, a, b, c are supposed to be pairwise relatively prime. Therefore $c = 1$, and the three numbers are $c = 1, b = 2, a = 3$.

Problem 2.83 Given that a and b are both prime numbers and $p = a^b + b^a$ is also prime. What is p?
Solution: 17.
If a and b are both odd primes, $p = a^b + b^a$ is an even number greater than 2, not a prime. So a or b should be 2. Assume $a = 2$ without loss of generality, and let $b = 2k + 1$. So $p = 2^{2k+1} + (2k+1)^2 = 2 \cdot 4^k + b^2$. If $b \neq 3$, $b^2 \equiv 1 \pmod 3$, and so $2 \cdot 4^k + b^2 \equiv 2 + 1 \equiv 0 \pmod 3$, and clearly $p > 3$, so p is not a prime, a contradiction. Therefore $b = 3$. At the end, $p = a^b + b^a = 2^3 + 3^2 = 17$.

Problem 2.84 Let n be a positive integer, and $\dfrac{n(n+1)}{2} - 1$ is a prime number. Find all possible values of n.

Solution: 2 or 3.

$\dfrac{n(n+1)}{2} - 1 = \dfrac{(n+2)(n-1)}{2}$. If $n = 1$ this expression is 0. If $n = 2$, the expression is 2, a prime; if $n = 3$, the expression is 5, a prime. For $n \geq 4$, the pair $n+2$ and $n-1$ consists of an even number at least 4 and an odd number at least 3, therefore the expression is not a prime. Therefore the only solutions are 2 and 3.

Problem 2.85 Given four cards with red, yellow, white, and blue colors, each card having a digit on it. Mike put the cards in a row in the order of red, yellow, white, and blue, to form a four digit number. Then he calculated the difference between this four digit number and 10 times the sum of its digits. He found out that no matter what digit was on the white card, the result of the calculation was always 1998. What are the digits on the red, yellow, and blue cards?

Solution: 2, 1, 8.

Let the digits on the red, yellow, white, and blue cards be r, y, w, b respectively. So $1000r + 100y + 10w + b - 10(r + y + w + b) = 1998$. Simplifying to get $990r + 90y - 9b = 1998$. Dividing by 9, $110r + 10y - b = 222$. Comparing the digits of both sides, get $b = 8, y = 1, r = 2$.

Problem 2.86 In a mathemagic show, the mathemagician asked Nick (a person he picked from the audience) to (1) think about a three digit number $\overline{abc}$; and (2) write down five numbers: $\overline{acb}, \overline{bac}, \overline{bca}, \overline{cab}, \overline{cba}$; and (3) add up these five numbers to get N. As soon as Nick said the value of N, the mathemagician announced the original number $\overline{abc}$. If $N = 3194$, what was $\overline{abc}$?

Solution: 358.

$3194 + \overline{abc} = \overline{acb} + \overline{bac} + \overline{bca} + \overline{cab} + \overline{cba} + \overline{abc} = 222(a + b + c)$, so we examine the multiples of 222 that are greater than 3194. $222 \times 15 = 3330$, but $3330 = 3194 + 136$, and the sum of digits of 136 is not 15. $222 \times 16 = 3552$, and $3552 = 3194 + 358$, and the sum of digits of 358 is 16, which is exactly right. No other numbers work. Thus $\overline{abc} = 358$.

Problem 2.87 From natural numbers $1, 2, 3, \ldots, 1000$, at most how many can be selected such that the sum of any three of the selected numbers is a multiple of 18?

Solution: 56.

Any two of the selected numbers have the same remainder mod 18, and the remainder has to be 0, 6, or 12. Choose $6, 24, 42, \ldots, 996$.

Problem 2.88 Find positive integer n that is divisible by both 5 and 49 and has exactly 10 positive divisors.

Solution: $12005 = 5 \times 7^4$.

To have 10 positive divisors, the number should be in the form pq^4 or p^9, where p and q are prime. In this case, the number n is already divisible by the primes 5 and 7, and the exponent of 7 is at least 2, thus it must be 5×7^4.

Problem 2.89 Let N be the least common multiple of $1, 2, 3, \ldots, 1998, 1999, 2000$, and 2^k be the maximum power of 2 that divides N. What is k?

Solution: 10.

It is the highest power of 2 under 2000, and $2^{10} = 1024$, $2^{11} = 2048$, therefore $k = 10$.

Problem 2.90 Partition the first n positive integers into several non-intersecting subsets, so that none of the subsets contain both m and $2m$ for any m. At least how many subsets should there be?

Solution: 2.

Write everything as $2^k a$, where a is odd. One set contains all for even k, one contains odd k.

Problem 2.91 Find all three digit number n such that the remainder when n is divided by 11 is equal to the sum of the squares of n's digits.

Solution: 100 and 101.

Let the three digit number be $\overline{abc}$. Since $a^2 + b^2 + c^2 < 11$, Each digit is at most 3. List all possible numbers (at most 21 of them) and check.

Problem 2.92 Attach a positive integer N to the right of any positive integer (for example, attaching 8 to the right of 57, we get 578), if the new number is always divisible by N no matter what the other positive integer is, then call N a "magic number". Find all "magic numbers" less than 2000.

Solution: The numbers are: 1,2,5,10,20,25,50, 100, 125, 200, 250, 500, 1000, 1250, totally 14 of them.

Categorize according to the number of digits. We show that k digit "magic numbers" is always a factor of 10^k: indeed, if N is a k-digit number, and E is any number, then $E \times 10^k + N = mN$ for some integer m. Therefore $E \times 10^k = (m-1)N$. Because E is arbitrary, this means N must be a factor of 10^k. For $k = 1$, N=1,2,5. For $k = 2$, $N = 10, 20, 25, 50$. For $k = 3$, $N = 100, 125, 200, 250, 500$. For $k = 4$ and under 2000, $N = 1000, 1250$.

Problem 2.93 Given three cards, each with an integer between 1 and 10 (inclusive) on it. After shuffling, deal the three cards to Adam, Bob, and Chris, one card each. Everyone write down the number on his card, and repeat the process: shuffle, deal, record. After some rounds, each person adds up the numbers he received. The sums are

---

13, 15, and 23. What are the numbers on the three cards?

Solution: 3, 5, 9.

The total is 51, so there are 3 rounds, and sum of the three numbers is 17. List the 15 possibilities that three positive integers between 1 and 10 add up to 17, then examine each case.

Problem 2.94 Is it possible to express 99^{99} as the sum of 99 consecutive odd positive integers? How about 99! ?

Solution: Yes for 99^{99}, no for 99!.

$99^{99} = (99^{98} - 98) + (99^{98} - 96) + \cdots + (99^{98} + 98)$. 99! is an even number.

Problem 2.95 From the numbers $1, 2, 3, \ldots, 999$, cross out the least possible number of numbers so that none of the remaining numbers is the product of two other remaining numbers. Which numbers should be crossed out?

Solution: $2, 3, \ldots, 30, 31$.

To show that removing 30 numbers is necessary, The triples $(2, 61, 2 \times 61), (3, 60, 3 \times 60), \ldots, (31, 32, 31 \times 32)$ are all distinct, and removing fewer than 30 numbers will leave one of the triples intact.

Problem 2.96 Is there a 3 digit number $\overline{abc}$ such that $\overline{abc} = \overline{ab} + \overline{bc} + \overline{ac}$?

Solution: No.

Assume possible, then $100a + 10b + c = 10a + b + 10b + c + 10a + c$, which simplifies to $80a = b + c$, which is impossible since a, b, c are all single digits.

Problem 2.97 Given any 17-digit number, reverse its digits to get another number, and then add the new number and the original number. Show that at least one digit in the sum is even.

Solution: Assume all digits in the sum are odd. Let the 17-digit number be $\overline{ab\ldots cd}$. Reversing its digits, we get $\overline{dc\ldots ba}$. From the units digits, $a + d$ is odd; thus from the highest digits, $a + d$ is unchanged, so $c + d$ (with possible carry from the lower digit) does not have carry. Thus the last two digits, $cd + ba$, does not carry over to the hundreds digit. Therefore we can remove the first two and last two digits, ab and cd, and the resulting 13-digit number still have the property: the sum of the number and its digit-reversal have all digits odd. This process can be repeated until only one digit left. For a single digit, the reversal is itself, and it is not possible to have an odd sum. Contradiction. Therefore, at least one digit in the original sum is even.

Problem 2.98 A magic coin machine behaves as follows. If you put in a penny, it returns a dime and a nickel. If you put in a nickel, it returns 4 dimes. If you put in a dime, it returns 3 pennies. Becky started with a penny and a nickel, and kept putting

coins into the machine and collected the returned coins. Is it possible at some point of time that the number of pennies Becky has is exactly 10 less than the number of dimes?
Solution: No.
The sum of the numbers of dimes and pennies is always odd. So the difference between them is also odd. Thus it is not possible to have a difference of 10.

Problem 2.99 The 3×3 table below contains 9 primes numbers. Define an "operation" as adding the same positive integer to the 3 numbers in one row or one column. Is it possible to change all numbers in the table to the same number after several operations?

2	3	5
13	11	7
17	19	23

Solution: No.
The sum of the 9 numbers is 100, which is 1 mod 3. Each operation doesn't change the remainder mod 3. And if all 9 numbers are the same, the sum is 0 mod 3, which is impossible to reach.

Problem 2.100 Find the sum of all the digits in the numbers $1, 2, 3, \ldots, 9999999$.
Solution: 315000000.
Pair up 0 and 9999999, 1 and 9999998, etc. In each pair, the sum of each corresponding digit is 9, so the digital sum of each pair is 63. There are totally 5000000 pairs, and therefore the answer is $5000000 \times 63 = 315000000$.

Problem 2.101 A department store distributes 9999 raffle tickets to the customers, each ticket has a 4-digit number from 0001 to 9999. If the sum of the first two digit equals the sum of the last two digits, then the ticket is called a "lucky ticket". For example, ticket number 0945 is a lucky ticket. Show that the sum of all the numbers on the lucky tickets is divisible by 101.
Solution: Pair up k and $9999 - k$; the sum of each pair is 9999, which is a multiple of 101.

Problem 2.102 Let
$$\frac{m}{n} = 1 + \frac{1}{2} + \frac{1}{3} + \cdots + \frac{1}{88}$$
where $\gcd(m, n) = 1$. Show that $89 \mid m$.
Solution: Pair up the first and last terms in the sum, and the second and second-to-last terms, and so on. Since $1 + \frac{1}{88} = \frac{89}{88}, \frac{1}{2} + \frac{1}{87} = \frac{89}{2 \times 87}$, and so on, each pair has

numerator 89. Since 89 is a prime, the common denominator does not have factor 89, therefore in the final result $\dfrac{m}{n}$, m is a multiple of 89.

Problem 2.103 One integer n equals the sum of 4 distinct fractions of form $\dfrac{m}{m+1}$ (m is a positive integer). Find this integer n, and also find at least one such set of 4 fractions that add up to n.

<u>Solution</u>: $n = 3$, and the sets are $\left\{\dfrac{1}{2},\dfrac{2}{3},\dfrac{6}{7},\dfrac{41}{42}\right\}$, $\left\{\dfrac{1}{2},\dfrac{2}{3},\dfrac{7}{8},\dfrac{23}{24}\right\}$, $\left\{\dfrac{1}{2},\dfrac{2}{3},\dfrac{8}{9},\dfrac{17}{18}\right\}$, $\left\{\dfrac{1}{2},\dfrac{2}{3},\dfrac{9}{10},\dfrac{14}{15}\right\}$, $\left\{\dfrac{1}{2},\dfrac{3}{4},\dfrac{4}{5},\dfrac{19}{20}\right\}$, and $\left\{\dfrac{1}{2},\dfrac{3}{4},\dfrac{5}{6},\dfrac{11}{12}\right\}$.

Since $\dfrac{1}{2} \le \dfrac{m}{m+1} < 1$, the sum of 4 distinct such fractions is strictly between 2 and 4, and the only possible integer value is 3.

To find such sets of fractions, consider the difference between each fraction and 1, which is of the form $\dfrac{1}{m+1}$. We shall find 4 distinct positive integers $a < b < c < d$ such that

$$\dfrac{1}{a}+\dfrac{1}{b}+\dfrac{1}{c}+\dfrac{1}{d}=1.$$

a has to be 2, since $\dfrac{1}{3}+\dfrac{1}{4}+\dfrac{1}{5}+\dfrac{1}{6}=\dfrac{19}{20}<1.$

b can take values 3 or 4 or 5, since $\dfrac{1}{6}+\dfrac{1}{7}+\dfrac{1}{8}<\dfrac{1}{2}.$

If $a=2, b=3$, $\dfrac{1}{6}=\dfrac{1}{c}+\dfrac{1}{d}$, which is $cd-6c-6d=0$. Completing the rectangle, $(c-6)(d-6)=36$. For integers $6<c<d$, we get solutions $(c,d)=(7,42)$ or $(8,24),(9,18),(10,15).$

If $a=2, b=4$, $\dfrac{1}{4}=\dfrac{1}{c}+\dfrac{1}{d}$. Solutions $(c,d)=(5,20)$ or $(6,12).$

If $a=2, d=5$, $\dfrac{3}{10}=\dfrac{1}{c}+\dfrac{1}{d}$. No integer solutions $5<c<d$ exists.

Finally, for each set of solutions (a,b,c,d), construct the 4 fractions

$$\left\{\dfrac{a-1}{a},\dfrac{b-1}{b},\dfrac{c-1}{c},\dfrac{d-1}{d}\right\}.$$

Problem 2.104 Find the largest n such that $n!$ has exactly 106 zeros at the end.
<u>Solution</u>: 434.
Use estimation. The number of zeros at the end of $n!$ is determined by the number of factor 5 (each factor 5 matched with a factor 2 produces a 0 at the end, and there are more factors of 2 than 5). Given that the number $100!$ has 24 zeros, it is estimated that n

is about 430. There are 86 multiples of 5 between 1 and 430. There are 17 multiples of 25, and 3 multiples of 125 between 1 and 430. Totally, there are $86 + 17 + 3 = 106$ of factor 5, which gives 106 zeros. Four additional numbers, 431, 432, 433, and 434 do not contribute more zeros at the end, so the largest number is 434.

Problem 2.105 A book has 192 pages and is printed double-sided on 96 sheets of paper. Each page has its page number printed at a corner. Kiran tore 25 sheets out of the book, and added up all the page numbers printed on these sheets. Is it possible that the sum is 2010?

Solution: No.

The two page numbers on each sheet are consecutive integers, whose sum is odd. The sum of 25 odd numbers is odd, and it cannot equal 2010.

Problem 2.106 The Gauss Middle School has 98 students, each has a unique student number, from 1 through 98. Is it possible to let the students stand in several rows, such that there is a student in each row whose number equals the sum of the numbers of the rest of the students in the same row?

Solution: No.

If the required seating arrangement were possible, then the sum of the student numbers in each row is an even number, so the total sum of the student numbers should be even. However, the sum of all the 98 numbers is an odd number: $1 + 2 + 3 + \cdots + 98 = \dfrac{98(98 + 1)}{2} = 49 \times 99$. Contradiction.

Problem 2.107 Ninety-nine students participated in the Planetary Math Olympiad. There are 30 questions in the PMO, and the scoring is as follows. There are 15 base points; add 5 for each correct answer, subtract 1 for each incorrect answer, and add 1 for each unanswered question. If all the scores were added up, is the sum an even number or an odd number?

Solution: Odd.

For each student, the base points is odd, and with 30 questions, the score for each question is odd, so the final score is odd.

Adding up 99 odd scores, the result is still odd.

Problem 2.108 Seventy-seven coins are put on the table, showing heads. First turn over all 77 coins. The second step, turn over 76 of them. The third step, turn over 75 of them, and so on. The 77th step, only turn over 1 of the coins. Is it possible to make all 77 coins show tails? If not, explain why. If yes, describe how it can be done.

Solution: Yes.

Use the pairing method. The first step turns every coins over. The second step and the

77th step are paired up to turn over all the coins, and so on.

Problem 2.109 The difference between two positive integers is multiplied by their product; can the final product be 45045?
Solution: No.
45045 is an odd number, and it can only be the product of odd numbers. The two positive integers are both odd, then their difference is even.

Problem 2.110 A certain 4-digit number satisfy the following: its tens digit minus 1 equals its units digit; the units digit plus 2 equals the hundreds digit; and if the digits of this 4-digit number is reversed, the new number plus the original number equals 9878. Find the original 4-digit number.
Solution: 1987.

Problem 2.111 Let a, b, c, d be a permutation of the numbers $1, 2, 3, 4$, satisfying $a < b, b > c, c < d$, and $\overline{abcd}$ is a 4-digit number. Find all such 4-digit numbers.
Solution: 1324, 1423, 2314, 2413, 3412.

Problem 2.112 Let n be the smallest multiple of 75 that has exactly 75 factors. Find $\dfrac{n}{75}$.
Solution: 432.
Using the formula for the number of factors, $75 = 5 \times 5 \times 3$, there could be many choices, and $n = 2^4 \times 3^4 \times 5^2$ turns out to be the smallest. So $\dfrac{n}{75} = 2^4 \times 3^3 = 432$.

Problem 2.113 Maggy has a deck of 100 cards. She starts with the card on top, and do the following: throw away the top card, and put the next top card to the bottom; then throw away the new top card, and put the next top card at the bottom, and so on, until only one card is left. Which card from the original deck is the remaining card?
Solution: The 72nd card.
After trying some small number of cards, the following pattern is found: If $N = 2^n$, then the remaining card is the last one, the 2^n-th. If $N = 2^n + m (m < 2^n)$, then the remaining card is the $2m$-th.
Second Solution: It is easy to see that if $N = 2^n$, then the remaining card is the last one, the 2^n-th. The highest power of 2 under 100 is 64. To reach 64 cards, we need to throw away 36 cards. Each step we throw away one card and move one card to the bottom, and after 36 such steps, the 72nd card is moved to the bottom. At this point there are 64 cards left in the deck, so the bottom card, the 72nd one from the original deck, is the one left at the end.

Problem 2.114 What is the largest even number that cannot be written as the sum of two odd composite numbers?

Solution: 38.

The odd composite numbers are (in increasing order): 9, 15, 21, 25, 27, 33, 35, ...

It is easy to see that 38 cannot be written as the sum of these numbers. To show that 38 is the largest, we demonstrate that any even number greater than 38 can be expressed as the sum of odd composite numbers. To start, $40 = 25 + 15$, $42 = 33 + 9$, and $44 = 35 + 9$. Each of these sums contains a multiple of 3. For any even numbers above 44, its difference from one of $\{20, 42, 44\}$ is a multiple of 3. Thus any even number beyond 38 is the sum of two odd composite numbers. Thus 38 is the largest.

Problem 2.115 (AIME 1983) Find the largest two-digit prime factor for the integer $\binom{200}{100}$.

Solution: 61.

$\binom{200}{100} = \dfrac{199 \times 198 \times \cdots \times 102 \times 101}{100!}$. Let this largest two-digit prime factor be p, then it occurs as a factor one more time on the numerator than the denominator. To be the largest, it should appear twice on top and once on the bottom. Thus $3p < 200$. So $p < 67$. The largest prime number below 67 is 61, and this is what we want.

Problem 2.116 (Kiev 1973) Find three primes numbers whose product is five times their sum.

Solution: 2,5,7.

Let p, q, r be the three primes, then $pqr = 5(p + q + r)$. Clearly one of the primes is 5. Let $p = 5$, and then $5qr = 5(5 + q + r)$, so $qr - q - r = 5$. Completing the rectangle, $(q - 1)(r - 1) = 6$. Since $q - 1$ and $r - 1$ are all integers, we get the only set of prime solutions: $p = 5, q = 2, r = 7$.

Problem 2.117 (Leningrad 1980) Let p and q be primes. The equation $x^4 - px^3 + q = 0$ has an integer root. Find the values of p and q.

Solution: $p = 3, q = 2$.

By the Rational Root theorem, the integer root is a factor of q, so it is one of the following: ± 1 and $\pm q$. For negative values of x, $x^4 - px^3 + q > 0$, so the root is positive. If $x = 1$ is a root, $1 - p + q = 0$, so $p = q + 1$. The only two primes satisfying this requirement is $q = 2, p = 3$.

If $x = q$ is a root, $q^4 - pq^3 + q = 0$, thus $q^3 - pq^2 + 1 = 0$, ad $1 = pq^2 - q^3 = q^2(p - q)$, which is not possible.

Thus the answer is $p = 3 q = 2$.

Problem 2.118 (Kiev 1978) Find the smallest positive integers a and $b(b > 1)$, such that $\sqrt{a\sqrt{a\sqrt{a}}} = b$.
Solution: $a = 256, b = 128$.
$a^7 = b^8$, so b must be the 7th power of some number >1. To make it the smallest, $b = 2^7$ and then $a = 2^8$.

Problem 2.119 Arrange the numbers $1, 2, 3, \ldots, 999$ on a circle, in that order. Start from 1, do the following: skip 1, cross out 2 and 3; skip 4, cross out 5 and 6. Each step skip one number and cross out the next two. Which number is the last one remaining?
Solution: 406.
If there are 3^k numbers (k is an integer), the number 1 is the last one remaining. 999 is not a power of 3. The largest power of 3 below 999 is $3^6 = 729$. If we cross out 270 numbers, there are 729 numbers remaining. To cross out 270 numbers, we need 135 steps. After 135 steps, 135 numbers are skipped and 270 numbers are crossed out, so we move over 405 numbers, and there are 729 numbers left. And the next number to visit is 406, which will be the last one remaining.

Problem 2.120 Form 4-digit numbers with the digits $0, 1, 2, 3, 4$, with no repeating digits within each number (for example, 1023, 3412). Find the sum of all such 4-digit numbers.
Solution: 259980.
If we remove the requirement of 4-digit numbers (that is, we allow the first digits to be 0), then there are $5 \times 4 \times 3 \times 2 = 120$ numbers. To find the sum of these numbers, notice that each digit occurs at each place the same number of times among these 120 numbers, so the sum is $(0 + 1 + 2 + 3 + 4) \times 120/5 \times 1111 = 266640$.
Now we subtract the sum of the numbers with 0 as the first digit. There are $4 \times 3 \times 2 = 60$ such numbers, and the sum is $(1 + 2 + 3 + 4) \times 24/4 \times 111 = 6660$.
The final sum is $26640 - 6660 = 259980$.

Problem 2.121 Twenty-seven countries send delegations to an international conference, each country has two representatives. Is it possible to arrange the 54 people around a round table, so that between the two people from any country, there are 9 people from other countries?
Solution: No. Use mod 54; Assume 1 and 11 are from the same country. Then 11 and 21 are not from the same country, and 21 and 31 are from the same country, and so on; $20k + 1$ and $20k + 11$ are from the same country; if $k = 13$, we get 261 and 271, which are equivalent to 45 and 1, are from the same country, contradiction.

Problem 2.122 A magic square is a square matrix with the property that the sums of the numbers on each row, column, and diagonal are the same. This sum is called the "magic sum". Is it possible that a 3×3 magic square has a magic sum 1999?

Solution: No.

Let the magic square be

$$
\begin{array}{ccc}
a & b & c \\
d & e & f \\
g & h & i
\end{array}
$$

and let the magic sum be k. Then $a+b+c = d+e+f = g+h+i = a+e+i = c+e+g = b+e+h = k$, and $(d+e+f)+(b+e+h)+(a+e+i)+(c+e+g)-(a+b+c)-(d+e+f)-(g+h+i) = 3e = k$. This means that k is a multiple of 3. The number 1999 is not a multiple of 3, so it is not a magic sum.

Problem 2.123 Express the fractions $\dfrac{7}{332}$ and $\dfrac{1949}{1999}$ in the form $\dfrac{1}{m}+\dfrac{1}{n}$, where m,n are positive integers. If not possible, explain why.

Solution: $\dfrac{7}{332} = \dfrac{1}{48}+\dfrac{1}{3984}$, and no solution for $\dfrac{1949}{1999}$.

For $\dfrac{7}{332}$: Factor $332 = 2^2 \times 83$. The numerator 7 has a multiple 84 that is $83+1$, thus we can work it out as follows:

$$
\frac{7}{332} = \frac{7}{2^2 \times 83} = \frac{7\times 12}{2^2 \times 83 \times 12} = \frac{83+1}{2^2 \times 83 \times 12} = \frac{83}{2^2 \times 83 \times 12} + \frac{1}{2^2 \times 83 \times 12}
$$
$$
= \frac{1}{48}+\frac{1}{3984}.
$$

For $\dfrac{1949}{1999}$: The fractions that can be of this form are (in decreasing order): $\dfrac{1}{2}+\dfrac{1}{2}=1$, $\dfrac{1}{2}+\dfrac{1}{3}=\dfrac{5}{6}$, etc. The number $\dfrac{1949}{1999}$ is between 1 and $\dfrac{5}{6}$, so it is not possible.

Problem 2.124 A five-digit number N consists of 5 distinct nonzero digits, and N equals the sum of all possible 3-digit numbers made up of 3 of its 5 digits. Find all such 5-digit numbers N.

Solution: 35964.

$N = \overline{abcde}$. Based on the requirement, the sum of all possible 3-digit numbers made up of 3 of the 5 digits of N is the same as N, so $N = (a+b+c+d+e) \times 12 \times 111$, thus $9 \mid N$, and then $9 \mid (a+b+c+d+e)$, therefore $(12\times 111 \times 9) \mid N$. Also, $a+b+c+d+e$ is at least $1+2+3+4+5 = 15$ and at most $5+6+7+8+9 = 35$, it must be either 18 or 27. Only 27 satisfies the requirements.

Problem 2.125 Evaluate:

$$\left\lfloor \frac{199 \times 1}{97} \right\rfloor + \left\lfloor \frac{199 \times 2}{97} \right\rfloor + \cdots + \left\lfloor \frac{199 \times 96}{97} \right\rfloor.$$

Solution: $9504 \ (= 198 \times 48)$.

Pair up $\left\lfloor \dfrac{199 \times 1}{97} \right\rfloor$ and $\left\lfloor \dfrac{199 \times 96}{97} \right\rfloor$. Since $\dfrac{199 \times 1}{97} + \dfrac{199 \times 96}{97} = 199$, and each of the

two fractions is not an integer, we get $\left\lfloor \dfrac{199 \times 1}{97} \right\rfloor + \left\lfloor \dfrac{199 \times 96}{97} \right\rfloor = 198$. Similarly, the

second term and second-to-last term add up to 198, and so do all other pairs. Totally there are 48 pairs.

Problem 2.126 Given plenty of apples, pears, and oranges, all mixed together in one pile. You want to separate the fruits into several piles, each containing a random number of all three kinds of fruits, to guarantee that no matter how many fruits each pile have, you can always select two piles that, when combined together, there are even number of each kind of fruits in the combined pile. How many piles should you separate the fruits into?

Solution: 9.

There are 8 possible parity distributions of three fruits: (even, even, even), (even, even, odd), (even, odd, even), (even, odd, odd), (odd, even, even), (odd, even, odd), (odd, odd, even), (odd, odd, odd). Making 9 piles guarantees that two piles have identical parity for each fruit, and combining those piles creates one pile that has an even number of each kind of fruit.

Problem 2.127 Can the number 1010 be expressed as the sum of 10 consecutive integers?

Solution: No.

The 10 consecutive integers can be expressed as $x - 4, x - 3, x - 2, x - 1, x, x + 1, x + 2, x + 3, x + 4, x + 5$. Their sum is $10x + 5$, an odd number, and it cannot equal an even number.

Problem 2.128 How many even numbers are among the first 100 Fibonacci numbers?

Solution: 33.

The Fibonacci numbers are $1, 1, 2, 3, 5, 8, 13, 21, 34, \ldots$

The parity pattern is odd, odd, even, odd, odd, even, $\ldots$.

That is one even number at each third term. So among the first 100 Fibonacci numbers, there are 33 even numbers.

Problem 2.129 Assume that in a chess game, the winner earns 1 point, the loser gets

-1, and both get 0 if it is a draw. In a tournament among several students, each student played one game against every other student. Given that one of the students received a total of 7 points, and another student received 20 points. Show that there was at least one draw during the games.

Solution: If there are no draws, then the score of each game for each student is odd. Since everyone played the same number of games, their total scores should all have the same parity. The scores 7 and 20 cannot exist simultaneously.

Problem 2.130 There are the 909 numbers on the board, $1, 2, \ldots, 909$. Each step, erase any two numbers from the board and write their nonnegative difference onto the board, until there is only one number left. Is this last number even or odd?

Solution: Odd.

For integers x and y, $x + y = (x - y) + 2y$, so the difference of two numbers has the same parity as the sum of the same two numbers. Therefore, each step, the new number written on board has the same parity as the sum of the two erase the numbers. Hence the last number on the board has the same parity as the total sum of all the original numbers.

We calculate $1 + 2 + 3 + \cdots + 909 = \dfrac{909 \times (909 + 1)}{2} = 909 \times 455$ is an odd number, so the final answer is odd.

Problem 2.131 (Putnam 1989) Let K be the set of all positive integers consisting of alternating digits 1 and 0: $\{1, 101, 10101, 1010101, \ldots\}$. Which elements of K are prime numbers?

Solution: Only 101 is prime.

1 is not a prime; if there are more than 3 digits, the numbers can be expressed as $\dfrac{10^{2n+2} - 1}{10^2 - 1} = \dfrac{10^{n+1} - 1}{10 - 1} \cdot \dfrac{10^{n+1} + 1}{10 + 1} = \dfrac{111 \cdots 1 \cdot (10^{n+1} + 1)}{10 + 1}$. Whether n is even or odd, the top is divisible by 11.

3. Algebra

Problem 3.1 Factor $10x^2y^2 - 15xy^3 + 25xy^2z$.
Solution: $5xy^2(2x - 3y + 5z)$.

Problem 3.2 Factor $6x(a-b)^4 - 30x(b-a)^3$.
Solution: $6x(a-b)^3(a-b+5)$.

Problem 3.3 Factor the following:

(a) $-2x^{5n-1}y^n + 4x^{3n-1}y^{n+2} - 2x^{n-1}y^{n+4}$
    Solution: $-2x^{n-1}y^n(x^n - y)^2(x^n + y)^2$

(b) $x^3 - 8y^3 - z^3 - 6xyz$
    Solution: $(x - 2y - z)(x^2 + 4y^2 + z^2 + 2xy + xz - 2yz)$

(c) $a^2 + b^2 + c^2 - 2bc + 2ca - 2ab$
    Solution: $(a - b + c)^2$

(d) $a^7 - a^5b^2 + a^2b^5 - b^7$
    Solution: $= (a+b)^2(a-b)(a^4 - a^3b + a^2b^2 - ab^3 + b^4)$

Problem 3.4 Factor $a^{32} - b^{32}$.
Solution: $(a^{16} + b^{16})(a^8 + b^8)(a^4 + b^4)(a^2 + b^2)(a+b)(a-b)$.

Problem 3.5 Factor $x^{15}+x^{14}+x^{13}+\cdots+x^2+x+1$.
Solution: $(x^8+1)(x^4+1)(x^2+1)(x+1)$.

$$x^{15}+x^{14}+x^{13}+\cdots+x^2+x+1 = \frac{x^{16}-1}{x-1}$$
$$= \frac{(x^8+1)(x^4+1)(x^2+1)(x+1)(x-1)}{x-1}$$
$$= (x^8+1)(x^4+1)(x^2+1)(x+1).$$

Problem 3.6 Factor the following:
(a) $2x^2-7x+3$.
  Solution: $(2x-1)(x-3)$.

(b) $3k^2-5k-2$.
  Solution: $(3k+1)(k-2)$.

(c) $2p^2+p-3$.
  Solution: $(2p+3)(p-1)$.

Problem 3.7 Factor the following:
(a) $x^9+x^6+x^3-3$.
  Solution: $(x-1)(x^2+x+1)(x^6+2x^3+3)$.

$$x^9+x^6+x^3-3 = (x^9-1)+(x^6-1)+(x^3-1)$$
$$= (x^3-1)(x^6+x^3+1)+(x^3-1)(x^3+1)+(x^3-1)$$
$$= (x^3-1)(x^6+2x^3+3)$$
$$= (x-1)(x^2+x+1)(x^6+2x^3+3)$$

(b) $(m^2-1)(n^2-1)+4mn$.
  Solution: $(mn+m-n+1)(mn-m+n+1)$.

$$(m^2-1)(n^2-1)+4mn = m^2n^2-n^2-m^2+1+4mn$$
$$= m^2n^2+2mn+1-(n^2-2mn+m^2)$$
$$= (mn+1)^2-(n-m)^2$$
$$= (mn+m-n+1)(mn-m+n+1)$$

(c) $(x+1)^4+(x^2-1)^2+(x-1)^4$.
  Solution: $(3x^2+1)(x^2+3)$. Let $a=x+1, b=x-1$, then the expression is $a^4+a^2b^2+b^4=a^4+2ab+b^4-a^2b^2=(a^2+b^2)^2-(ab)^2=(a^2+ab+b^2)(a^2-ab+b^2)$. Change back to x, and simplify, we get $(3x^2+1)(x^2+3)$.

Second Solution: We can also expand everything and simplify first. $(x+1)^4 + (x^2-1)^2 + (x-1)^4 = 3x^4 + 10x^2 + 3$ and at this point it is easy to factor as a quadratic polynomial in x^2.

(d) $a^3b - ab^3 + a^2 + b^2 + 1$.
   Solution: $(a^2 - ab + 1)(b^2 + ab + 1)$.

$$
\begin{aligned}
& a^3b - ab^3 + a^2 + b^2 + 1 \\
=\ & a^3b - ab^3 + a^2 + b^2 + 1 + a^2b^2 + ab - a^2b^2 - ab \\
=\ & (a^2b^2 - ab^3 + b^2) + (a^3b - a^2b^2 + ab) + (a^2 - ab + 1) \\
=\ & (a^2 - ab + 1)b^2 + (a^2 - ab + 1)ab + (a^2 - ab + 1) \\
=\ & (a^2 - ab + 1)(b^2 + ab + 1)
\end{aligned}
$$

Problem 3.8 For the equation $x^2 - 402x + k = 0$, one of the roots plus three equals 80 times the other root. Find the value of k.
Solution: $k = 1985$.
Let the two roots be x_1 and x_2, then $x_1 + x_2 = 402$ and $x_1x_2 = k$. Also $x_1 + 3 = 80x_2$. Solve for x_1 and x_2, we get $x_1 = 397$ and $x_2 = 5$. Therefore $k = x_1x_2 = 1985$.

Problem 3.9 Without solving the equation, find the number of real roots for x: $(n^2 + 1)x^2 - 2nx + (n^2 + 4) = 0$.
Solution: $\Delta = (2n)^2 - 4(n^2 + 1)(n^2 + 4) = 4n^2 - 4n^4 - 20n - 16 = -4(n^2 + 4)^2(n^2 + 2)^2 < 0$. There are no real roots.

Problem 3.10 For what values of m the equation $4x^2 + 8x + m = 0$ has two distinct real roots?
Solution: $m < 4$.
$\Delta = 8^2 - 4 \cdot 4m = 64 - 16m > 0$, so $m < 4$.

Problem 3.11 A quadratic equation has two roots $\dfrac{2}{3}$ and $-\dfrac{1}{2}$, what is this equation? (multiple answers are possible)
Solution: $6x^2 - x - 2 = 0$ or any equation with coefficients proportional to this one.
Let the equation be $ax^2 + bx + c = 0$, then by Vieta's formulas,
$$\frac{2}{3} + \left(-\frac{1}{2}\right) = \frac{1}{6} = -\frac{b}{a}$$
$$\frac{2}{3} \cdot \left(-\frac{1}{2}\right) = -\frac{1}{3} = \frac{c}{a}.$$
Thus the equation is $x^2 - \dfrac{1}{6}x - \dfrac{1}{3} = 0$, which can be re-written as $6x^2 - x - 2 = 0$. (Or any equation with coefficients proportional to this one.)

Problem 3.12 Two real numbers have sum -10 and product -5, find these two numbers.

Solution: $-5 + \sqrt{30}$ and $-5 - \sqrt{30}$.

These two numbers are the roots of the equation

$$x^2 + 10x - 5 = 0.$$

Solving to get $x = \dfrac{-10 \pm \sqrt{10^2 + 4(5)}}{2} = -5 \pm \sqrt{30}$. So the two numbers are $-5 + \sqrt{30}$ and $-5 - \sqrt{30}$.

Problem 3.13 Let a, b, c be real numbers. The following three quadratic equations each has only one real root (i.e. double roots):

$$
\begin{aligned}
ax^2 + 2bx + c &= 0 \\
bx^2 + 2cx + a &= 0 \\
cx^2 + 2ax + b &= 0
\end{aligned}
$$

Show that $a = b = c$.

Solution: Discriminants all equal 0, so

$$
\begin{aligned}
4b^2 - 4ac &= 0 \\
4c^2 - 4ba &= 0 \\
4a^2 - 4cb &= 0
\end{aligned}
$$

Canceling the 4, and add all three together, $a^2 + b^2 + c^2 - ac - ba - cb = 0$. Completing the squares, $\dfrac{1}{2}((a-b)^2 + (b-c)^2 + (c-a)^2) = 0$. This means $a = b = c$.

Problem 3.14 Given that $x^2 + 2px + 1 = 0$ has two real roots, one is greater than 1, and the other is less than 1. What is the range of the possible values of p?

Solution: $p < -1$.

The product of the roots is 1 (positive), and one is greater than 1, so both are positive. Thus $x_1 + x_2 > 0$, $-2p > 0$, $p < 0$. In addition, $\Delta = 4p^2 - 4 > 0$, thus $p < -1$ or $p > 1$, so the final answer is $p < -1$.

Problem 3.15 For equation $2x^2 + mx - 2m + 1 = 0$, the sum of squares of the two real roots is $\dfrac{29}{4}$. Find the value of m.

Solution: $m = 3$.

Use Vieta's Theorem, $x_1^2 + x_2^2 = (x_1 + x_2)^2 - 2x_1x_2 = \left(\dfrac{m}{2}\right)^2 - (-2m+1) = \dfrac{29}{4}$. Solve for m to get $m = 3$ and $m = -11$. The other solution $m = -11$ doesn't provide real roots for the equation, so throw away.

Problem 3.16 Let x_1 and x_2 be the two roots of the equation $4x^2 - 8x + k = 0$. Given that $\dfrac{1}{x_1} + \dfrac{1}{x_2} = \dfrac{8}{3}$, find k.

Solution: $k = 3$.

By Vieta's formulas, $x_1 + x_2 = 2, x_1 x_2 = \dfrac{k}{4}$. Thus

$$\frac{1}{x_1} + \frac{1}{x_2} = \frac{x_1 + x_2}{x_1 x_2} = \frac{8}{k} = \frac{8}{3}.$$

Therefore $k = 3$.

Problem 3.17 If x_1 and x_2 are the two real roots of $x^2 + x + q = 0$, and $|x_1 - x_2| = q$, find the value of q.

Solution: $q = \sqrt{5} - 2$.

Note that $q^2 = |x_1 - x_2|^2 = (x_1 + x_2)^2 - 4x_1 x_2 = 1 - 4q$. Solve to find $q = -2 \pm \sqrt{5}$. Since q is an absolute value, take the positive root for q.

Problem 3.18 The equation $x^2 + (a - 6)x + a = 0$ has two integer roots. Find the value of a.

Solution: 0 or 16.

$x_1 + x_2 = 6 - a$ and $x_1 x_2 = a$, so $x_1 + x_2 + x_1 x_2 + 1 = 7$. Factoring, $(x_1 + 1)(x_2 + 1) = 7$. Since 7 is a prime, there are two possibilities: $x_1 + 1 = 1, x_2 + 1 = 7$ OR $x_1 + 1 = -1, x_2 + 1 = -7$. From these we get the two roots are 0 and 6, or -8 and -2. The number a is the product of the two roots, so the possible values for a are 0 and 16.

Problem 3.19 Given equation in x: $x^2 + 2mx + m + 2 = 0$.

(a) For what values of m does the equation have two positive roots?

Solution: $-2 < m \leq -1$.

First the discriminant $\Delta = 4m^2 - 4(m + 2) \geq 0$, which is $m^2 - m - 2 \geq 0$, whose solution is $m \in (-\infty, -1] \cup [2, +\infty)$.

$x_1 > 0$ and $x_2 > 0$ is equivalent to the system of inequalities $x_1 + x_2 > 0$ and $x_1 x_2 > 0$. By Vieta's formulas, $-2m > 0$ and $m + 2 > 0$, and the solution is $m \in (-2, 0)$.

Therefore the final answer is $m \in (-2, -1]$.

(b) For what values of m does the equation have one positive root and one negative root?

Solution: $m < -2$.

From the discriminant, $\Delta = 4m^2 - 4(m + 2) \geq 0$, which is $m^2 - m - 2 \geq 0$, and the solution is $m \in (-\infty, -1] \cup [2, +\infty)$.

If $x_1 > 0$ and $x_2 < 0$, $x_1 x_2 < 0$, by Vieta's formulas, $m + 2 < 0$. Thus $m \in$

$(-infty, -2)$.
So the final answer is $m \in (-\infty, -2)$.

Problem 3.20 The quadratic equation $x^2 + 2kx + 2k^2 - 1 = 0$ has at least one negative root. Find the possible range of values for k.

Solution: $-\dfrac{\sqrt{2}}{2} < k \leq 1$.

First, the equation has real roots, so the discriminant $\Delta = 4k^2 - 4(2k^2 - 1) \geq 0$, which has solution $-1 \leq k \leq 1$. This is the largest possible range for k.

To find the range of k that the equation has at least one negative root, we find the opposite first: for what values of k the given equation has two non-negative roots? This question translates into $x_1 \geq 0$ and $x_2 \geq 0$, which is equivalent to $x_1 + x_2 \geq 0$ and $x_1 x_2 \geq 0$. Thus, $-2k \geq 0$ and $2k^2 - 1 \geq 0$, and then $k \leq -\dfrac{\sqrt{2}}{2}$. This means the equation has two non-negative roots if $-1 \leq k \leq -\dfrac{\sqrt{2}}{2}$.

Therefore, to have at least one negative root, k should be in the range $-\dfrac{\sqrt{2}}{2} < k \leq 1$.

Problem 3.21 Find the real solution to the system of equations: $x + y = 2$ and $xy - z^2 = 1$.

Solution: $x = 1, y = 1, z = 0$.

We use Vieta's formulas backwards. Since x, y, z are real, and $x + y = 2$ and $xy = z^2 - 1$, x and y are the two real roots of the equation $t^2 - 2t + (z^2 + 1) = 0$, $\Delta \geq 0$ means $4 - 4(z^2 + 1) \geq 0$, so $z^2 \leq 0$, which requires $z = 0$. Consequently, the two roots for t are both 1, so $x = y = 1$.

Problem 3.22 Given $p > 0$ and $q < 0$, how many positive roots does the equation $x^2 + px + q = 0$ have?

Solution: 1.

$\Delta = p^2 - 4q > 0$ means there are two real roots, and $x_1 x_2 = q < 0$ means one is positive and one is negative.

Problem 3.23 The sum of squares of the roots of equation $x^2 + 2kx = 3$ is 10. Find the possible values of k.

Solution: ± 1.

The equation can be written as $x^2 + 2kx - 3 = 0$. We are given that $x_1^2 + x_2^2 = 10$. By Vieta's formulas, $x_1^2 + x_2^2 = (x_1 + x_2)^2 - 2x_1 x_2 = 4k^2 + 6 = 10$, so $k^2 = 1$, which gives $k = \pm 1$.

Problem 3.24 If x_1 and x_2 are integer roots of equation $x^2 + mx + 2 - n = 0$, and $(x_1^2 + 1)(x_2^2 + 1) = 10$, how many possible pairs (m, n) are there?
Solution: The answer is 6.
Since x_1 and x_2 are integers, the possible values for $x_1^2 + 1$ and $x_2^2 + 1$ are $1, 10$ and $2, 5$. In the first case, $x_1 = 0, x_2 = \pm 3$; and in the second case, $x_1 = \pm 1$ and $x_2 = \pm 2$. The order of x_1 and x_2 does not matter, since we only need to find the possible pairs of (m, n). By Vieta's formulas, $-m = x_1 + x_2, 2 - n = x_1 x_2$, so there are 6 different pairs for (m, n): $(3, 2), (-3, 2), (3, 0), (1, 4), (-1, 4), (-3, 0)$.

Problem 3.25 For the equation $x^2 + mx + n = 0$, the difference between the two roots is p and the product of the two roots is q. What is $m^2 + n^2$ in terms of p and q?
Solution: $p^2 + 4q + q^2$.
It is given that $x_1 - x_2 = p$ and $x_1 x_2 = q$. Also by Vieta's formulas, $x_1 + x_2 = -m, x_1 x_2 = n$, thus $n = q$ and $m^2 + n^2 = (x_1 + x_2)^2 + (x_1 x_2)^2 = (x_1 - x_2)^2 + 4x_1 x_2 + (x_1 x_2)^2 = p^2 + 4q + q^2$.

Problem 3.26 Let x_1, x_2 be two positive integer roots of equation $x^2 + px + 1997 = 0$. Find the value of $\dfrac{p}{(x_1 + 1)(x_2 + 1)}$.
Solution: $-1/2$.
By Vieta's formulas, $x_1 + x_2 = -p$ and $x_1 x_2 = 1997$. Since 1997 is a prime, and x_1, x_2 are positive integers, they must be 1 and 1997. Then $x_1 + x_2 = 1998$ and $p = -1998$. So
$$\frac{p}{(x_1 + 1)(x_2 + 1)} = \frac{-1998}{x_1 x_2 + x_1 + x_2 + 1} = \frac{-1998}{1997 + 1998 + 1} = -\frac{1}{2}.$$

Problem 3.27 The two real roots of $x^2 + (m - 2)x + 5 - m = 0$ are both greater than 2. Find the possible range of values for real number m.
Solution: $-5 < m \le -4$.
Since there are two real roots, the discriminant $\Delta = (m - 2)^2 - 4(5 - m) \ge 0$, which simplifies to $m^2 - 16 \ge 0$ and thus $m \le -4$ or $m \ge 4$.
Given $x_1 > 2$ and $x_2 > 2$, it means $x_1 - 2 > 0$ and $x_2 - 2 > 0$, and this is equivalent to $(x_1 - 2) + (x_2 - 2) > 0$ and $(x_1 - 2)(x_2 - 2) > 0$. (Note: it is NOT equivalent to $x_1 + x_2 > 4$ and $x_1 x_2 > 4$. Consider the example $x_1 = 1, x_2 = 5$, not both greater than 2, but satisfy the inequalities $x_1 + x_2 > 4$ and $x_1 x_2 > 4$.)
By Viet's formulas, $x_1 + x_2 = 2 - m, x_1 x_2 = 5 - m$, thus $(x_1 - 2) + (x_2 - 2) = 2 - m - 4 = -m - 2 > 0$ and $(x_1 - 2)(x_2 - 2) = x_1 x_2 - 2(x_1 + x_2) + 4 = 5 - m - 2(2 - m) + 4 = m + 5 > 0$, so $-5 < m < -2$.
Combine the results, the final answer is $-5 < m \le -4$.

Problem 3.28 Let m be integer. The equation $x^2 + mx - m + 1 = 0$ has two distinct positive integer roots. Find m.

Solution: -5.

$x_1 + x_2 = -m$ and $x_1 x_2 = -m + 1$, therefore $x_1 x_2 - x_1 - x_2 = 1$. Since x_1 and x_2 are two distinct positive integers, we complete the rectangle: $x_1 x_2 - x_1 - x_2 + 1 = 2$, and $(x_1 - 1)(x_2 - 1) = 2$. Thus $x_1 - 1 = 1$ and $x_2 - 1 = 2$, or $x_1 - 1 = -1$ and $x_2 - 1 = -2$ (impossible). Hence, $x_1 = 2$ and $x_2 = 3$. Consequently $m = -(x_1 + x_2) = -5$. Verification: For $m = -5$, the equation is $x_2 - 5x + 6 = 0$, which has roots 2 and 3.

Problem 3.29 Solve the equation $\dfrac{3-x}{2+x} = 5 - \dfrac{4(2+x)}{3-x}$.

Solution: $1/2$ and -1.

Let $y = \dfrac{3-x}{2+x}$, then $y = 5 - \dfrac{4}{y}$. This gives a quadratic equation in y; $y^2 - 5y + 4 = 0$, and $y = 1$ or $y = 4$. Set $\dfrac{3-x}{2+x} = 1$ and $\dfrac{3-x}{2+x} = 4$, and solve for x to get $x = \dfrac{1}{2}$ and $x = -1$.

Problem 3.30 Solve: $\dfrac{15}{x+1} = \dfrac{15}{x} - \dfrac{1}{2}$.

Solution: -6 and 5.

$30x = 30(x+1) - x(x+1)$, so $x^2 + x - 30 = 0$, and this has roots $x = -6$ and $x = 5$. Both check out fine.

Problem 3.31 Solve: $\dfrac{x-3}{x+1} - \dfrac{x+1}{3-x} = \dfrac{5}{2}$.

Solution: 7 and -5.

Let $y = \dfrac{x-3}{x+1}$. Then $y + \dfrac{1}{y} = \dfrac{5}{2}$, and so $y = \dfrac{1}{2}$ or $y = \dfrac{1}{2}$. These give solutions $x = 7$ and $x = -5$. Both are roots after verifying.

Problem 3.32 Solve: $\left(\dfrac{x+1}{x^2-1}\right)^2 - 4\left(\dfrac{x+1}{x^2-1}\right) + 3 = 0$.

Solution: 2 and $4/3$.

$x = -1$ is not a root, so cancel it and simplify: $\left(\dfrac{1}{x-1}\right)^2 - \dfrac{4}{x-1} + 3 = 0$. Then let $y = \dfrac{1}{x-1}$. Solve and get $y = 1$ and $y = 3$, so $x = 2$ and $x = 4/3$. Both are roots after verifying.

Problem 3.33 Solve: $\dfrac{3x-1}{x^2+1} - \dfrac{3x^2+3}{3x-1} = 2$.

Solution: 0 and -3.

Let $y = \dfrac{3x-1}{x^2+1}$. So $y - \dfrac{3}{y} = 2$, which simplifies to $y^2 - 2y - 3 = 0$. Then $y = 3$ (no real roots for x) and $y = -1$, we get $x = 0$ and $x = -3$. Verifying, both are roots.

Problem 3.34 Solve: $2x^4 - 9x^3 + 14x^2 - 9x + 2 = 0$.
Solution: $x = 1$(double root), $x = 2$, and $x = 1/2$.

Divide everything by x^2: $2\left(x^2 + \dfrac{1}{x^2}\right) - 9\left(x + \dfrac{1}{x}\right) + 14 = 0$. Let $y = x + \dfrac{1}{x}$, then $y^2 = x^2 + \dfrac{1}{x^2} + 2$, so $2(y^2 - 2) - 9y + 14 = 0$, and $y = 2$ and $y = 5/2$. Then solve for x and get $x = 1$ (double root), $x = 2$, and $x = 1/2$.

Second Solution: We can recognize that the sum of all coefficients equals 0, which implies $x = 1$ is a root, and $(x-1)$ is a factor of the left hand side, and factor the left hand side. Then we notice $(x-1)$ is another factor, and the remaining factor is quadratic, which is easy to factor again.

Problem 3.35 Find all solutions to $|||x+1| - 1| - 1| = 1$.
Solution: $-4, -2, 0, 2$.
Case (1) If $||x+1| - 1| - 1 = 1$, then $||x+1| - 1| = 2$. Case (1a) $|x+1| - 1 = 2$, $|x+1| = 3$, so $x = 2$ or $x = -4$. Case (1b) $|x+1| - 1 = -2$ impossible.
Case (2) If $||x+1| - 1| - 1 = -1$, then $||x+1| - 1| = 0$, so $|x+1| = 1$, thus $x = 0$ or $x = -2$.
Therefore there are 4 solutions: $-4, -2, 0, 2$.

Problem 3.36 Solve: $x^2 - \sqrt{3x^2 + 7} = 1$.
Solution: $\pm\sqrt{6}$.
Multiply 3 on both sides, $3x^2 - 3\sqrt{3x^2 + 7} = 3$. Let $y = \sqrt{3x^2 + 7}$. Then $y^2 - 7 - 3y = 3$, and $y^2 - 3y - 10 = 0$, so $y = 5$ and $y = -2$ (negative value for y is throw away). so $x = \pm\sqrt{6}$.

Problem 3.37 Solve: $2x^2 - \sqrt{4x^2 - 12x} = 6x + 4$.
Solution: 4 and -1.
Simplify to get $x^2 - \sqrt{x^2 - 3x} = 3x + 2$, and $x^2 - 3x - \sqrt{x^2 - 3x} - 2 = 0$, so let $y = \sqrt{x^2 - 3x}$. Then $y^2 - y - 2 = 0$, thus $y = 2$ (negative value is thrown away.) Solving for x, and $x = 4$ and $x = -1$.

Problem 3.38 Solve: $\sqrt{x^2 + 3x + 7} - \sqrt{x^2 + 3x - 9} = 2$.
Solution: 3 and -6.
Let $y = \sqrt{x^2 + 3x + 7}$, then $y - \sqrt{y^2 - 16} = 2$, then $y^2 - 4y + 4 = y^2 - 16$, so $y = 5$.

Solve for x to get $x = 3$ and $x = -6$, both check out alright.

Second Solution: We can also use two new variables: let $u = \sqrt{x^2 + 3x + 7}$, and $v = \sqrt{x^2 + 3x - 9}$, then $u - v = 2$, and also $u^2 - v^2 = 16$. After division, $u + v = 8$, therefore $u = 5, v = 3$. Solving $\sqrt{x^2 + 3x + 7} = 5$, we get $x = 3$ and $x = -6$.

Problem 3.39 Solve for x:

$$x^2 - \sqrt{x^2 - 3x + 5} = 3x + 1.$$

Solution: 4 and -1.

The equation is rewritten as $x^2 - 3x + 5 - \sqrt{x^2 - 3x + 5} - 6 = 0$.

Use change of variables: $y = \sqrt{x^2 - 3x + 5}$. Then $y^2 - y - 6 = 0$, and $y = 3$, or $y = -2$ (extraneous). From $y = 3$, we get $x = 4$ or -1.

Problem 3.40 Let p, q be positive integers, and the roots of $px^2 - qx + 1985 = 0$ are both prime numbers. What is the value of $12p^2 + q$?

Solution: 414.

By Vieta's formulas, $x_1 + x_2 = q/p$, and $x_1 x_2 = 1985/p$. Thus $1985/p$ is an integer and is a product of two primes. We know that $1985 = 5 \times 397$ where 5 and 397 are primes, so $p = 1$ and the two roots are 5 and 397. Therefore $q = 5 + 397 = 402$, and $12p^2 + q = 12 + 402 = 414$.

Problem 3.41 Let A, B, p be integers, and the two roots of $x^2 + px + 19 = 0$ are both exactly 1 greater than the two roots of $x^2 + Ax + B = 0$, respectively. Find the value of $B - A$.

Solution: 18.

Let u, v be the two roots of $x^2 + px + 19 = 0$, and w, t be the two roots of $x^2 + Ax + B = 0$. Then $u = w + 1$ and $v = t + 1$.

By Vieta's formulas, $u + v = -p$ and $uv = 19$, $w + t = -A$ and $wt = B$.

Thus $19 = (w + 1)(t + 1) = wt + w + t + 1 = B - A + 1$, therefore $B - A = 18$.

Problem 3.42 If a, b are integers, and equation $ax^2 + bx + 1 = 0$ has two distinct positive roots, both less than 1. What is the minimum possible value of a?

Solution: 5.

The equation has two distinct real roots, so $\Delta = b^2 - 4a > 0$.

Since $0 < x_1 < 1, 0 < x_2 < 1$, by Vieta's theorem, $x_1 + x_2 = b/a > 0$, $x_1 x_2 = 1/a < 1$, thus $a > 1$, and $b < 0$.

We are given that a and b are both integers, so we try small values for a to see if there are two distinct roots in $(0, 1)$.

(1) $a = 2$. So $2x^2 + bx + 1 = 0$, and $x = \dfrac{-b \pm \sqrt{b^2 - 8}}{4} < 1$. Solving the inequality for b to get $-3 < b < 0$. However, this does not give real roots for the equation.

(2) $a = 3$. In this case $3x^2 + bx + 1 = 0$, and $x = \dfrac{-b \pm \sqrt{b^2 - 12}}{6} < 1$. Solving the inequality for b to get $-4 < b < 0$. This does not give real roots for the equation either.

(3) $a = 4$. In this case $4x^2 + bx + 1 = 0$, and $x = \dfrac{-b \pm \sqrt{b^2 - 16}}{8} < 1$. Solving the inequality for b to get $-5 < b < 0$. Only when $b = -4$ this equation has real root $-1/2$, and it is a double root, not two distinct real roots as required.

(4) $a = 5$. In this case $5x^2 + bx + 1 = 0$, and $x = \dfrac{-b \pm \sqrt{b^2 - 20}}{10} < 1$. Solving the inequality for b to get $-6 < b < 0$. Here $b = -5$ works nicely, so we found the minimum a that satisfies the requirements.

Problem 3.43 The equation $2kx^2 + (8k+1)x + 8k = 0$ has two distinct real roots for x. Find the range of values for k.
Solution: $k > -1/16$ and $k \neq 0$.
This is a quadratic equation, so $2k \neq 0$. Also $\Delta = (8k+1)^2 - 4(2k)(8k) > 0$, which simplifies to $16k + 1 > 0$. and then $k > -\dfrac{1}{16}$. Therefore the final answer is $\left(-\dfrac{1}{16}, 0\right) \cup (0, +\infty)$.

Problem 3.44 For what values of a, b the equation $x^2 + 2(1+a)x + (3a^2 + 4ab + 4b^2 + 2) = 0$ has real roots?
Solution: $a = 1, b = -1/2$.
$\Delta = -4(2a^2 - 2a + 4b^2 + 4ab + 1) = -4[(a-1)^2 + (a+2b)^2] \geq 0$, then we must have $a = 1, b = -1/2$.

Problem 3.45 The equation $|x^2 - 5x| = a$ has exactly two distinct real roots. What is the possible range of values for a?
Solution: $a = 0$ or $a > 25/4$.
$a = 0$ is obviously a good value.
If $a > 0$, the two equations $x^2 - 5x - a = 0$ and $x^2 - 5x + a = 0$ combined have two distinct roots, that means the second equation should have no real roots. Therefore, $5^2 + 4a > 0$ and $5^2 - 4a < 0$, which lead to $a > -\dfrac{25}{4}$ and $a > \dfrac{25}{4}$, thus $a > \dfrac{25}{4}$.

Problem 3.46 Let a be a rational number. Suppose the roots of the equation $x^2 + 3(a-1)x + (2a^2 + a + b) = 0$ are always rational numbers, no matter what rational values a takes. What is the value of b?

Solution: $b = -28$.

$\Delta = a^2 - 22a + 9 - 4b$, as an expression in a, should be a perfect square, which means the equation $a^2 - 22a + 9 - 4b = 0$ in a has a double root, and $\Delta_2 = 22^2 - 4 \times (9 - 4b) = 0$, solve for b.

Problem 3.47 Given an equation in x: $x^2 + (m-2)x + \frac{1}{2}m - 3 = 0$.

(a) Show that no matter what real value m takes, the equation always has two distinct real roots.

    Solution: The discriminant $\Delta = (m-3)^2 + 7$, which is always positive.

(b) Let x_1, x_2 be the two real roots of the given equation, and assume they also satisfy $2x_1 + x_2 = m + 1$, find the value of m.

    Solution: 0 or $\frac{17}{12}$.

    By Vieta's theorem, $x_1 + x_2 = -m + 2$ and $x_1 x_2 = \frac{1}{2}m - 3$. Combine with $2x_1 + x_2 = m + 1$ and solve for m, we get two possible solutions: $m = 0$ or $m = \frac{17}{12}$.

Problem 3.48 Given a quadratic equation in x: $x^2 - 2(m-2)x + m^2 = 0$. Does there exist a real number m such that this equation has two real roots and the sum of squares of these two roots equals 56? If so, find the value of m; if not, explain why.

Solution: Yes, $m = -2$.

$x_1^2 + x_2^2 = 56$, and use Vieta's theorem to get $m = -2$ or 10. Since the discriminant $\Delta \geq 1$, we get $m \leq 1$, thus $m = -2$.

Problem 3.49 Given that $x = \dfrac{1}{\sqrt{3}-2}, y = \dfrac{1}{\sqrt{3}+2}$, evaluate $\dfrac{x^2 + xy + y^2}{x+y}$.

Solution: $-\dfrac{13\sqrt{3}}{6}$.

Rationalize the denominators of x and y: $x = -2 - \sqrt{3}$ and $y = 2 - \sqrt{3}$. Thus $x + y = -2\sqrt{3}$ and $xy = -1$. The numerator $x^2 + xy + y^2 = (x+y)^2 - xy = 12 + 1 = 13$. Therefore the answer is $\dfrac{13}{-2\sqrt{3}} = -\dfrac{13\sqrt{3}}{6}$.

Problem 3.50 Given that $\dfrac{x^2}{x^2-2} = \dfrac{1}{1-\sqrt{2}-\sqrt{3}}$, evaluate

$$\dfrac{\dfrac{1}{1-x} - \dfrac{1}{1+x}}{\dfrac{x}{x^2-1} + x}.$$

Solution: $-(\sqrt{2}+\sqrt{3})$.

From given, $\dfrac{x^2-2}{x^2}=1-\sqrt{2}-\sqrt{3}$. Thus, $\dfrac{2}{x^2}=\sqrt{2}+\sqrt{3}$. Now we simplify the expression:

$$
\dfrac{\dfrac{1}{1-x}-\dfrac{1}{1+x}}{\dfrac{x}{x^2-1}+x} = \dfrac{-(x+1)-(x-1)}{x+x(x^2-1)} \quad \text{(multiplying } x^2-1 \text{ on top and bottom)}
$$

$$
= \dfrac{-2x}{x^3}
$$

$$
= -\dfrac{2}{x^2}
$$

$$
= -(\sqrt{2}+\sqrt{3}).
$$

Problem 3.51 Real numbers x and y satisfy $|2x-y+1|+2\sqrt{3x-2y+4}=0$. Find the value of

$$
1-\dfrac{x-y}{x-2y}\div\dfrac{x^2-y^2}{x^2-4xy+4y^2}.
$$

Solution: $\dfrac{15}{7}$.

$|2x-y+1|\geq 0$ and $\sqrt{3x-2y+4}\geq 0$, therefore they must be both equal to 0. Thus $2x-y+1=0$ and $3x-2y+4=0$, so we solve and get $x=2,y=5$. To evaluate the expression, we simply first:

$$
1-\dfrac{x-y}{x-2y}\div\dfrac{x^2-y^2}{x^2-4xy+4y^2} = 1-\dfrac{x-y}{x-2y}\div\dfrac{(x+y)(x-y)}{(x-2y)^2}
$$

$$
= 1-\dfrac{x-y}{x-2y}\cdot\dfrac{(x-2y)^2}{(x+y)(x-y)}
$$

$$
= 1-\dfrac{x-2y}{x+y}
$$

$$
= \dfrac{3y}{x+y}
$$

$$
= \dfrac{15}{7}.
$$

Problem 3.52 Solve the equation:

$$
\sqrt{1+\dfrac{2}{x-1}}-\sqrt{1-\dfrac{2}{x+1}}=\dfrac{3}{2}.
$$

Solution: $\dfrac{5}{3}$.

The equation can be transformed into

$$\sqrt{\frac{x+1}{x-1}} - \sqrt{\frac{x-1}{x+1}} = \frac{3}{2}.$$

Use the change of variables $y = \sqrt{\frac{x+1}{x-1}}$, and solve for y, we get $y = 2$ or $y = -\frac{1}{2}$. For $y = 2$, we get $x = \frac{5}{3}$. For $y = -\frac{1}{2}$, no solution. Verify, and $x = \frac{5}{3}$ is a solution.

Problem 3.53 Let a be a real number, and the equation $x^2 + a^2x + a = 0$ has real roots for x. Find the maximum possible root x.
Solution: $x_{\max} = \sqrt[3]{2}/2$.
See the equation as a quadratic equation in a, $xa^2 + a + x^2 = 0$.
Because a is a real number, the discriminant $1^2 - 4x^3 \geq 0$, so $4x^3 \leq 1$, thus $x \leq \frac{1}{\sqrt[3]{4}} = \frac{\sqrt[3]{2}}{2}$.

Problem 3.54 Solve for x: $(x - \sqrt{3})x(x+1) + 3 - \sqrt{3} = 0$.
Solution: $\sqrt{3} - 1$ and $\pm\sqrt[4]{3}$.
This is a cubic equation, and it is difficult to solve directly. So we let $y = \sqrt{3}$. Then $(x-y)x(x+1) + y^2 - y = 0$. Expand and factor the left hand side: $x^3 - x^2y + x^2 - xy + y^2 - y = (x - y + 1)(x^2 - y) = 0$, so we have $x = y - 1$ and $x^2 = y$. Thus the solutions: $x = \sqrt{3} - 1$ and $x = \pm\sqrt[4]{3}$.

Problem 3.55 Solve: $\sqrt{\sqrt{x+4}+4} = x$
Solution: $\frac{\sqrt{17}+1}{2}$.
First note that $x > 0$. Square both sides: $\sqrt{x+4} + 4 = x^2$.
Then $\sqrt{x+4} = x^2 - 4$, also note that $x > 2$.
Squaring again to get $x + 4 = x^4 - 2 \cdot 4x^2 + 4^2$.
To avoid factoring a quartic polynomial in x, we apply the following technique of "slave-as-master": who says x must be the variable and 4 has to be the constant? Treat the constant 4 as a variable, and x as a parameter. Then it is a quadratic equation in the constant "4":

$$4^2 - (2x^2 + 1)4 + (x^4 - x) = 0.$$

Using the quadratic formula on 4, we get

$$4 = \frac{2x^2 + 1 \pm \sqrt{(2x^2+1)^2 - 4(x^4 - x)}}{2} = \frac{2x^2 + 1 \pm (2x+1)}{2},$$

hence $4 = x^2 + x + 1$ or $4 = x^2 - x$. Solve each of these and throw away the negative roots, $x = \dfrac{\sqrt{13} - 1}{2}$ or $x = \dfrac{\sqrt{17} + 1}{2}$.

Further check shows that $x = \dfrac{\sqrt{13} - 1}{2} < 2$ is also extraneous.

So there is only one root $x = \dfrac{\sqrt{17} + 1}{2}$.

(It might help for easier understanding to do it this way: Let $y = 4$: $\sqrt{\sqrt{x+y}+y} = x$, and then solve for y in terms of x.)

Problem 3.56 Solve for x: $\sqrt{5 - \sqrt{5-x}} = x$.

Solution: $x = \dfrac{\sqrt{21} - 1}{2}$.

Same "slave-as-master" method as in Problem 3.55: let $y = 5$, then solve for y in terms of x. Note that $0 < x < \sqrt{5}$ to get rid of the extraneous roots.

Problem 3.57 Factor $(x^2 + 3x + 2)(4x^2 + 8x + 3) - 90$.

Solution: $(2x^2 + 5x + 12)(2x + 7)(x - 1)$.

$(x^2 + 3x + 2)(4x^2 + 8x + 3) - 90 = (x+1)(x+2)(2x+1)(2x+3) - 90 = [(x+1)(2x+3)][(x+2)(2x+1)] - 90 = (2x^2 + 5x + 3)(2x^2 + 5x + 2) - 90$.

Let $y = 2x^2 + 5x + 2$, then the original expression equals

$y(y+1) - 90 = y^2 + y - 90 = (y-9)(y+10) = (2x^2 + 5x - 7)(2x^2 + 5x + 12) = (2x + 7)(x - 1)(2x^2 - 5x + 12)$.

Problem 3.58 Factor $(x^2 + 4x + 8)^2 + 3x(x^2 + 4x + 8) + 2x^2$.

Solution: $(x+2)(x+4)(x^2 + 5x + 8)$.

Use a change of variables $y = x^2 + 4x + 8$, then

$$\begin{aligned}(x^2 + 4x + 8)^2 + 3x(x^2 + 4x + 8) + 2x^2 &= y^2 + 3xy + 2x^2 \\ &= (y + x)(y + 2x) \\ &= (x^2 + 5x + 8)(x^2 + 6x + 8) \\ &= (x^2 + 5x + 8)(x + 2)(x + 4).\end{aligned}$$

Problem 3.59 Factor $6x^4 + 7x^3 - 36x^2 - 7x + 6$.

Solution: $(2x + 1)(x - 2)(3x - 1)(x + 3)$.

$6x^4 + 7x^3 - 36x^2 - 7x + 6 = x^2\left(6x^2 + 7x - 36 - \dfrac{7}{x} + \dfrac{6}{x^2}\right)$.

Let $y = x - \dfrac{1}{x}$, $y^2 = x^2 - 2 + \dfrac{1}{x^2}$, thus

$$
\begin{aligned}
6x^4 + 7x^3 - 36x^2 - 7x + 6 &= x^2(6(y^2+2) + 7y - 36) \\
&= x^2(2y-3)(3y+8) \\
&= (2xy - 3x)(3xy + 8x) \\
&= (2x^2 - 3x - 2)(3x^2 + 8x - 3) \\
&= (2x+1)(x-2)(3x-1)(x+3).
\end{aligned}
$$

Problem 3.60 Factor $(x^2 + xy + y^2)^2 - 4xy(x^2 + y^2)$.
<u>Solution</u>: $(x^2 - xy + y^2)^2$.
Let $u = x^2 + y^2, v = xy$,

$$(x^2 + xy + y^2)^2 - 4xy(x^2 + y^2) = (u+v)^2 - 4uv = (u-v)^2 = (x^2 - xy + y^2)^2.$$

Problem 3.61 Let n be an integer, show that $n(n+1)(n+2)(n+3) + 1$ is a perfect square.
<u>Solution</u>: $n(n+1)(n+2)(n+3) + 1 = (n^2 + 3n + 1)^2$.

Problem 3.62 Factor $x^3 - 19x - 30$.
<u>Solution</u>: $(x+2)(x+3)(x-5)$.
Using the Rational Roots Theorem, check the factors of 30 and plug in those values for x. It is easy to check that $x = -2, -3, 5$ are the zeros of this polynomial.
So Based on the Factor Theorem, the factorization is $(x+2)(x+3)(x-5)$.

Problem 3.63 Show that $2x + 3$ is a factor of $2x^4 - 5x^3 - 10x^2 + 15x + 18$.
<u>Solution</u>: Plug in $x = -3/2$, and the value of the polynomial is 0. Thus $2x + 3$ is a factor by the Factor Theorem.

Problem 3.64 If k is an integer and $x^2 + 2kx - 3k^2$ has a factor $(x - 1)$, what should k be?
<u>Solution</u>: 1.
Plug in $x = 1$, the value of the polynomial should be 0. So $1 + 2k - 3k^2 = 0$. This is a quadratic equation in k, and $k = 1$ or $-1/3$. Since k is an integer, the answer is $k = 1$.

Problem 3.65 $x^4 + 5x^3 + 15x - 9$
<u>Solution</u>: $(x^2 + 3)(x^2 + 5x - 3)$.
Try the values $\pm 1, \pm 3, \pm 9$, none works. So we need to have two quadratic factors. Assume $x^4 + 5x^3 + 15x - 9 = (x^2 + ax + b)(x^2 + cx + d)$. To simplify matters, we split into three cases: (1) $b = 3, d = -3$; (2) $b = 1, d = -9$; (3) $b = 9, d = -1$. As long as one of the cases work, we succeed.

Case (1) $(x^2+ax+3)(x^2+cx-3) = x^4+5x^3+15x-9$. So $x^4+(a+c)x^3+acx^2+3(c-a)x-9 = x^4+5x^3+15x-9$.

Therefore, $a+c = 5$, $ac = 0$, $3(c-a) = 15$. Thus $a = 0, c = 5$. This already produces a solution, so there's no need to work on the other two cases.

Problem 3.66 $x^4-12x+323$
Solution: $(x^2-6x+17)(x^2+6x+19)$.
Use the method of undetermined coefficients: $x^4-12x+323 = (x^2+ax+17c)(x^2+bx+19c)$, where $c = \pm 1$.
The value $c = 1$ produces the solution.

Problem 3.67 x^3+3x^2-4
Solution: $(x-1)(x+2)^2$

Problem 3.68 $x^4-11x^2y^2+y^4$
Solution: $(x^2-3xy-y^2)(x^2+3xy-y^2)$

Problem 3.69 $x^3+9x^2+26x+24$
Solution: $(x+2)(x+3)(x+4)$

Problem 3.70 $(2x^2-3x+1)^2-22x^2+33x-1$
Solution: $x(x-3)(2x+3)(2x-3)$.
Let $y = 2x^2-3x$, then $(2x^2-3x+1)^2-22x^2+33x-1 = (y+1)^2-11y-1 = y^2-9y = y(y-9) = (2x^2-3x)(2x^2-3x-9) = x(x-3)(2x+3)(2x-3)$.

Problem 3.71 $x^4+7x^3+14x^2+7x+1$
Solution: $(x^2+3x+1)(x^2+4x+1)$.
The coefficients are symmetric, so let $y = x+\dfrac{1}{x}$, and it becomes quadratic in y.

Problem 3.72 $(x+3)(x^2-1)(x+5)-20$
Solution: $(x^2+4x-7)(x^2+4x+5)$.
$(x+3)(x^2-1)(x+5)-20 = (x+3)(x+1)(x-1)(x+5)-20 = (x^2+4x+3)(x^2+4x-5)-20$, then let $y = x^2+4x$.

Problem 3.73 $2x^2+3xy-9y^2+14x-3y+20$
Solution: $(2x-3y+4)(x+3y+5)$

Problem 3.74 $a^2+(a+1)^2+(a^2+a)^2$

Solution: $(a^2 + a + 1)^2$.
Expand the first two terms, $a^2 + (a+1)^2 = a^2 + a^2 + 2a + 1 = 2(a^2 + a) + 1$, so $a^2 + (a+1)^2 + (a^2+a)^2 = (a^2+a)^2 + 2(a^2+a) + 1 = (a^2+a+1)^2$.

Problem 3.75 $2acx + 4bcx + adx + 2bdx + 4acy + 8bcy + 2ady + 4bdy$
Solution: $(a+2b)(d+2c)(x+2y)$.

$$
\begin{aligned}
& (2acx+4bcx)+(adx+2bdx)+(4acy+8bcy)+(2ady+4bdy) \\
=\ & 2cx(a+2b)+dx(a+2b)+4cy(a+2b)+2dy(a+2b) \\
=\ & (a+2b)(2cx+dx+4cy+2dy) \\
=\ & (a+2b)((2c+d)x+(2c+d)(2y)) \\
=\ & (a+2b)(2c+d)(x+2y)
\end{aligned}
$$

Problem 3.76 $ab(c^2-d^2)-cd(a^2-b^2)$
Solution: Expand, then group. Answer: $(ac+bd)(bc-ad)$.

Problem 3.77 a^5+a+1.
Solution: $(a^2+a+1)(a^3-a^2+1)$.
Add and minus a^2: $a^5 - a^2 + a^2 + a + 1 = a^2(a^3-1)+(a^2+a+1) = (a^2+a+1)(a^3-a^2+1)$.

Problem 3.78 $x^4+y^4+z^4-2x^2y^2-2y^2z^2-2z^2x^2$
Solution: $(x+y+z)(x-y+z)(x+y-z)(x-y-z)$.

$$
\begin{aligned}
& x^4+y^4+z^4-2x^2y^2-2y^2z^2-2z^2x^2 \\
=\ & (x^4+y^4+z^4-2x^2y^2-2y^2z^2+2z^2x^2)-4z^2x^2 \\
=\ & (x^2-y^2+z^2)^2-4z^2x^2 \\
=\ & (x^2-y^2+z^2+2xz)(x^2-y^2+z^2-2xz) \\
=\ & ((x^2+2xz+z^2)-y^2)((x^2-2xz+z^2)-y^2) \\
=\ & ((x+z)^2-y^2)((x-z)^2-y^2) \\
=\ & (x+y+z)(x-y+z)(x+y-z)(x-y-z).
\end{aligned}
$$

Problem 3.79 $1+2a+3a^2+4a^3+5a^4+6a^5+5a^6+4a^7+3a^8+2a^9+a^{10}$.
Solution: $(a+1)^2(a^2+a+1)^2(a^2-a+1)^2$.
Grouping, this equals $(1+a+a^2+a^3+a^4+a^5)^2$, so answer is $(a+1)^2(a^2+a+1)^2(a^2-a+1)^2$.

Problem 3.80 $(x+1)(x+3)(x+5)(x+7)+15$
Solution: $(x+2)(x+6)(x^2+8x+10)$

Problem 3.81 Evaluate the following: $\dfrac{(1994^2 - 2000)(1994^2 + 3985) \times 1995}{1991 \cdot 1993 \cdot 1995 \cdot 1997}$.

Solution: 1996.

Let $x = 1994$, then the expression is

$$\frac{(x^2 - x - 6)(x^2 + 2x - 3)(x+1)}{(x-3)(x-1)(x+1)(x+3)} = \frac{(x+2)(x-3)(x+3)(x-1)(x+1)}{(x-3)(x-1)(x+1)(x+3)} = x+2 = 1996.$$

Problem 3.82 Distinguish the following concepts. Find examples for each of them.

(a) A zero-degree polynomial.

   Solution: A nonzero constant. Examples: 9, $-5/4$, etc.

(b) The zero polynomial.

   Solution: **0**. The value of this polynomial is 0 no matter what values the variables assume.

(c) Zero of a polynomial.

   Solution: A value for the variable where the polynomial's value equals 0. In other words, a root of the polynomial equation.

Solution: See the definitions.

Problem 3.83 For this problem, $f(x) = 3x + 2$, $g(x) = x - 7$, and $h(x) = x^2 - 4x + 4$. Compute the following values:

(a) $f(g(4))$

   Solution: -7

(b) $g(f(4))$

   Solution: 7

(c) $g(g(g(g(g(35)))))$

   Solution: 0

(d) $h(f(0))$

   Solution: 0

(e) $h(f(1))$

   Solution: 9

(f) $h(f(100))$

   Solution: 90000

(g) $f(g(1234567)) - g(f(1234567))$

   Solution: -14

Problem 3.84 In the polynomial $(7+x)(1+x^2)(5+x^4)(2+x^8)(3+x^{16})(10+x^{32})$, what is the coefficient of x^{54}?
Solution: 14.
Converted to binary, $54_{10} = 110110_2$. Thus the power $x^{54} = x^{32} \cdot x^{16} \cdot x^4 \cdot x^2$, and the coefficient is the product of the constant terms whose power of x does not appear, so the answer is $7 \cdot 2 = 14$.

Problem 3.85 The two roots of equation $x^2 + px + 1 = 0 (p > 0)$ has a difference 1. Find the value of p.
Solution: $\sqrt{5}$.

$$x_1 + x_2 = -p, x_1 x_2 = 1, x_1 - x_2 = 1, \text{ so } x_1 = \frac{1-\sqrt{5}}{2}, x_2 = \frac{-1-\sqrt{5}}{2}, \text{ so } p = \sqrt{5}.$$

Problem 3.86 Let $m \geq -1$ be a real number, and the equation $x^2 + 2(m-2)x + m^2 - 3m + 3 = 0$ has two distinct real roots x_1 and x_2. If $x_1^2 + x_2^2 = 6$, what is m?
Solution: $\dfrac{5 - \sqrt{17}}{2}$.
There are two distinct real roots, so $\Delta = 4(m-2)^2 - 4(m^2 - 3m + 3) = -4m + 4 > 0$. Therefore $m < 1$, and thus $-1 \leq m < 1$. So $6 = x_1^2 + x_2^2 = (x_1 + x_2)^2 - 2x_1 x_2 = 4(m-2)^2 - 2(m^2 - 3m + 3) = 2m^2 - 10m + 10$. Solve for m we get $m = \dfrac{5 \pm \sqrt{17}}{2}$. Given the $-1 \leq m < 1$ range, we get $m = \dfrac{5 - \sqrt{17}}{2}$.

Problem 3.87 Expand $(x^2 - x + 1)^6$ to get $a_{12}x^{12} + a_{11}x^{11} + \cdots + a_1 x + a_0$. Find the value of $a_{12} + a_{10} + a_8 + a_6 + a_4 + a_2 + a_0$.
Solution: 365.
Let $x = 1$:

$$1^6 = a_{12} + a_{11} + a_{10} + \cdots + a_2 + a_1 + a_0.$$

Let $x = -1$:

$$3^6 = a_{12} - a_{11} + a_{10} - \cdots + a_2 - a_1 + a_0.$$

Adding,

$$730 = 2(a_{12} + a_{10} + a_8 + a_6 + a_4 + a_2 + a_0),$$

So $a_{12} + a_{10} + a_8 + a_6 + a_4 + a_2 + a_0 = 365$.

Problem 3.88 Assume $(x-c)^2 \mid (4x^3 + 8x^2 - 11x + 3)$, find the value of c.
Solution: $1/2$.

Use the Rational Root Theorem to factor

$$4x^3 + 8x^2 - 11x + 3 = (2x-1)^2(x+3) = 4\left(x - \frac{1}{2}\right)^2(x+3).$$

Problem 3.89 Assume $(x-1)^2 \mid [x^4 + (m+n)x^3 + (m-n)x^2 + (m^2+2n-1)x + m+2]$. Find the value of m and n.
Solution: $m = 0, n = -1$ or $m = 1, n = -3$.
Use long division (or the Method of Undetermined Coefficients) to factor $f(x)$:

$$f(x) = (x-1)(x^3 + (m+n+1)x^2 + (2m+1)x - (m+2)).$$

Let $g(x) = x^3 + (m+n+1)x^2 + (2m+1)x - (m+2)$. By the Factor Theorem, $f(1) = 0$ and $g(1) = 0$, and we have equations

$$\begin{cases} 1 + m + n + m - n + m^2 + 2n - 1 + m + 2 &= 0, \\ 1 + m + n + 1 + 2m + 1 - (m+2) &= 0. \end{cases}$$

Solve and get $m = 0, n = -1$ or $m = 1, n = -3$.

Problem 3.90 Let $x = \dfrac{2}{2 + \sqrt{3} - \sqrt{5}}, y = \dfrac{2}{2 + \sqrt{3} + \sqrt{5}}$, evaluate:

$$\frac{x^4 y^4}{x^4 + y^4 + 6x^2 y^2 + 4x^3 y + 4xy^3}.$$

Solution: $97 - 56\sqrt{3}$.
Let $M = \dfrac{x^4 y^4}{x^4 + y^4 + 6x^2 y^2 + 4x^3 y + 4xy^3}$. Then

$$\frac{1}{M} = \frac{x^4 + y^4 + 6x^2 y^2 + 4x^3 y + 4xy^3}{x^4 y^4} = \frac{(x+y)^4}{x^4 y^4} = \left(\frac{1}{x} + \frac{1}{y}\right)^4.$$

Also $\dfrac{1}{x} + \dfrac{1}{y} = \dfrac{2 + \sqrt{3} - \sqrt{5}}{2} + \dfrac{2 + \sqrt{3} + \sqrt{5}}{2} = 2 + \sqrt{3}$.

Therefore $\dfrac{1}{M} = (2 + \sqrt{3})^4 = (7 + 4\sqrt{3})^2 = 97 + 56\sqrt{3}$, and then $M = 97 - 56\sqrt{3}$.

Problem 3.91 Distinct real numbers a and b satisfies $(a+1)^2 = 3 - 3(a+1), 3(b+1) = 3 - (b+1)^2$. Find the value of $b\sqrt{\dfrac{b}{a}} + a\sqrt{\dfrac{a}{b}}$.
Solution: -23.
a and b are the roots of an equation $(x+1)^2 + 3(x+1) - 3 = 0$, which is $x^2 + 5x + 1 = 0$.

The discriminant is positive. We have: $a + b = -5$, and $ab = 1$. Thus $b\sqrt{\dfrac{b}{a}} + a\sqrt{\dfrac{a}{b}} =$

$$-\frac{b}{a}\sqrt{ab} - \frac{a}{b}\sqrt{ab} = -\frac{a^2 + b^2}{ab}\sqrt{ab} = -\frac{(a+b)^2 - 2ab}{\sqrt{ab}} = -23.$$

Problem 3.92 Without solving the equation, find out the number of real roots of the following equation: $x^3 + 3x - 1 = 0$.
Solution: 1.
Use graphing argument. The graphs of $y = x^3$ and $y = 3x - 1$ has exactly one intersection.

Problem 3.93 Let m, p be positive integers. The two parabolas $P_1(x) = x^2 + 5x + m$ and $P_2(x) = x^2 + px + 2$ have no common points, and $P_1(100) < P_2(100)$. Find the value of $m + p$.
Solution: 6.
The fact that the two parabolas have no common points means that the equation $P_1(x) = P_2(x)$ has no solution. The equation $P_1(x) = P_2(x)$ is, in effect, $5x + m = px + 2$, a linear equation. It has no solution if $p = 5$ and $m \neq 2$. Given $p = 5$, and $P_1(100) < P_2(100)$, we have $m < 2$, and the only positive integer less than 2 is 1. Therefore $m = 1$ and $p = 5$, so $m + p = 6$.

Problem 3.94 An $l \times w \times h$ rectangular box has surface area 38 and volume 12. If $l + w + h = 8$, find the dimensions of the box.
Solution: $4 \times 3 \times 1$.
Since $lwh = 12$ and $2(lw + wh + hl) = 38$, the values l, w, h are the three roots of polynomial equation $x^3 - 8x^2 + 19x - 12 = 0$. Factor the polynomial, $(x-1)(x-3)(x-4) = 0$, so the three roots are $1, 3, 4$. Thus it is a $4 \times 3 \times 1$ box.

Problem 3.95 Let a, b, c, and d be the roots of $x^4 - 2x - 1990 = 0$. Find the value of $1/a + 1/b + 1/c + 1/d$.
Solution: $-1/995$.
$abc + bcd + cda + dab = 2$, and $abcd = -1990$, so $1/a + 1/b + 1/c + 1/d = (abc + bcd + cda + dab)/abcd = -1/995$.
Second Solution: Let $y = 1/x$, then $1 - 2y^3 - 1990y^4 = 0$, and the four roots for y are $1/a, 1/b, 1/c, 1/d$, so the answer is $-(-2)/(-1990) = -1/995$.

Problem 3.96 If a,b,c,d are four different numbers for which

$$\begin{cases} a^4 + a^2 + ka + 64 = 0 \\ b^4 + b^2 + kb + 64 = 0 \\ c^4 + c^2 + kc + 64 = 0 \\ d^4 + d^2 + kd + 64 = 0. \end{cases}$$

What is the value of $a^2 + b^2 + c^2 + d^2$?
Solution: -2.
a,b,c,d are roots of equation $x^4 + x^2 + kx + 64 = 0$ (Note that they can be complex roots).
So $a^2 + b^2 + c^2 + d^2 = (a+b+c+d)^2 - 2(ab+ac+ad+bc+bd+cd) = 0 - 2 \cdot 1 = -2$.

Problem 3.97 Find the sum of the 17th powers of the 17 roots of $x^{17} - 3x + 1 = 0$.
Solution: -17.
Each of the roots satisfy $x_i^{17} - 3x_i + 1 = 0$, thus $x_i^{17} = 3x_i - 1$. Using Vieta's formula, $\sum x_i = 0$, therefore
$$\sum x_i^{17} = \sum (3x_i - 1) = 3\sum x_i - 17 = -17.$$

Problem 3.98 Let x and y be nonzero real numbers satisfying $|x| + y = 3$ and $|x|y + x^3 = 0$, Find the value of $x + y$.
Solution: $4 - \sqrt{13}$.
substitute $y = 3 - |x|$ into $|x|y + x^3 = 0$, get $x^3 - x^2 + 3|x| = 0$. Case analysis for $x \geq 0$ and $x < 0$: If $x \geq 0$, $x^3 - x^2 + 3x = 0$, there are no nonzero real roots. If $x < 0$, $x^3 - x^2 - 3x = 0$, so $x = \dfrac{1 - \sqrt{13}}{2}$ (throw away the positive root), and $y = 3 + x$, so $x + y = 3 + 2x = 4 - \sqrt{13}$.

Problem 3.99 Find ordered pairs (x,y) of real numbers such that $x^2 - xy + y^2 = 13$ and $x - xy + y = -5$.
Solution: $(3,4),(4,3),(-2+\sqrt{3},-2-\sqrt{3}),(-2-\sqrt{3},-2+\sqrt{3})$.
Let $u = x+y$, $v = xy$, then $u^2 - 3v = 13$, and $u - v = -5$. Solve to get $(u,v) = (7,12)$ or $(u,v) = (-4,1)$. For each pair of (u,v), set up quadratic equation in t: $t^2 - ut + v = 0$ and x,y are the two roots. So $t^2 - 7t + 12 = 0$, or $t^2 + 4t + 1 = 0$. The solution for (x,y) are $(3,4),(4,3),(-2+\sqrt{3},-2-\sqrt{3}),(-2-\sqrt{3},-2+\sqrt{3})$.

Problem 3.100 If $x+y+z = 0$ and $x^3 + y^3 + z^3 = 288$, find the value of xyz.
Solution: 96.
An important factoring formula: $x^3 + y^3 + z^3 - 3xyz = (x+y+z)(x^2+y^2+z^2 - xy - yz - zx)$. Then $288 - 3xyz = 0$, so $xyz = 96$.

Problem 3.101 Let x be a real number such that $x^3 + 4x = 8$. Determine the value of

$x^7 + 64x^2$.

Solution: 128.

Use $x^3 = -4x + 8$ to reduce the exponents. $x^7 + 64x^2 = x(-4x+8)^2 + 64x^2 = 16x^3 - 64x^2 + 64x + 64x^2 = 16(x^3 + 4x) = 16 \times 8 = 128$.

Problem 3.102 The polynomial $p(x) = x^3 + 2x^2 - 5x + 1$ has three different roots a, b, and c. Find $a^3 + b^3 + c^3$.

Solution: -41.

$a+b+c = -2, ab+bc+ca = -5, abc = -1$. And $(a+b+c)^3 = a^3 + b^3 + c^3 + 3a^2b + 3ab^2 + 3b^2c + 3bc^2 + 3c^2a + 3ca^2 + 6abc = a^3 + b^3 + c^3 + 3(a+b+c)(ab+bc+ca) + abc$. So $a^3 + b^3 + c^3 = (-2)^3 - 3(-2)(-5) + (-1) = -41$.

Problem 3.103 Suppose that the roots of $3x^3 + 3x^2 + 4x - 11 = 0$ are a, b and c, and the roots of $x^3 + rx^2 + sx + t = 0$ are $a+b, b+c$, and $c+a$. Find t.

Solution: 5.

$t = -(a+b)(b+c)(c+a) = -(ab+bc+ca)(a+b+c) + abc = -(4/3)(-1) + (11/3) = 5$.

Problem 3.104 Let $P(x) = (x-1)(x-2)(x-3)$. For how many polynomials $Q(x)$ does there exist a polynomial $R(x)$ of degree 3 such that $P(Q(x)) = P(x)R(x)$?

Solution: 22.

The polynomial $P(Q(x))$ is of degree 6, so $Q(x)$ is quadratic. Let $Q(x) = ax^2 + bx + c$. $Q(x)$ is determined by values at three points. Given that $1, 2, 3$ are roots for $P(x) = 0$, we get that $P(Q(x)) = P(x)R(x) = 0$ if $x = 1, 2, 3$. Therefore $\{Q(1), Q(2), Q(3)\} \subset \{1, 2, 3\}$. There are $3^3 = 27$ possible value assignments for $(Q(1), Q(2), Q(3))$, and each determines one possible polynomial $Q(x)$ by the following system of equations.

$$\begin{aligned} a+b+c &= Q(1), \\ 4a+2b+c &= Q(2), \\ 9a+3b+c &= Q(3). \end{aligned}$$

But we don't want to include those choices that do not make quadratic polynomials. There are five such choices for $(Q(1), Q(2), Q(3))$: $(1,1,1)$, $(2,2,2)$, $(3,3,3)$, $(1,2,3)$, $(3,2,1)$. The first three choices give constant polynomials, and the last two are linear. So the final answer is 22.

Problem 3.105 Determine $x^2 + y^2 + z^2 + w^2$ if

$$\frac{x^2}{2^2 - 1^2} + \frac{y^2}{2^2 - 3^2} + \frac{z^2}{2^2 - 5^2} + \frac{w^2}{2^2 - 7^2} = 1,$$

$$\frac{x^2}{4^2 - 1^2} + \frac{y^2}{4^2 - 3^2} + \frac{z^2}{4^2 - 5^2} + \frac{w^2}{4^2 - 7^2} = 1,$$

$$\frac{x^2}{6^2 - 1^2} + \frac{y^2}{6^2 - 3^2} + \frac{z^2}{6^2 - 5^2} + \frac{w^2}{6^2 - 7^2} = 1,$$

$$\frac{x^2}{8^2 - 1^2} + \frac{y^2}{8^2 - 3^2} + \frac{z^2}{8^2 - 5^2} + \frac{w^2}{8^2 - 7^2} = 1.$$

Solution: 36.

$2^2, 4^2, 6^2, 8^2$ are the roots of the equation in t:

$$\frac{x^2}{t - 1^2} + \frac{y^2}{t - 3^2} + \frac{z^2}{t - 5^2} + \frac{w^2}{t - 7^2} = 1$$

which is

$$x^2(t - 3^2)(t - 5^2)(t - 7^2) + y^2(t - 1^2)(t - 5^2)(t - 7^2)$$
$$+ z^2(t - 1^2)(t - 3^2)(t - 7^2) + w^2(t - 1^2)(t - 3^2)(t - 5^2)$$
$$= (t - 1^2)(t - 3^2)(t - 5^2)(t - 7^2).$$

After some rearrangements, the equation becomes a quartic equation in t. By Vieta's Theorem,

$$2^2 + 4^2 + 6^2 + 8^2 = 1^2 + 3^2 + 5^2 + 7^2 + x^2 + y^2 + z^2 + w^2,$$

So $x^2 + y^2 + z^2 + w^2 = 36$.

4. Geometry

Problem 4.1 An equilateral triangle must also be equiangular, and vice versa. But this is not true for other polygons.

   (a) Give an example of an equiangular polygon that is not equilateral.
       <u>Solution</u>: Rectangle. Many examples are possible.
   (b) Give an example of an equilateral polygon that is not equiangular.
       <u>Solution</u>: Rhombus. Many examples are possible.

Problem 4.2 Complete the following table about polygons: name, sum of interior angles, sum of exterior angles, and measure of each angle in case of regular polygon. All angles are in degrees. Justify your answers.

# sides	Polygon	Int. angle sum	Ext. angle sum	Each angle (if regular)
3	Triangle	180	360	60
4	Quadrilateral	360	360	90
5	Pentagon	540	360	108
6	Hexagon	720	360	120
7	Heptagon	900	360	900/7
8	Octagon	1080	360	135
9	Nonagon	1260	360	140
10	Decagon	1440	360	144
12	Dodecagon	1800	360	150
20	Icosagon	3240	360	162

Solution: To justify the sum of interior angles of a triangle: In $\triangle ABC$, draw line through A that's parallel to $\overline{BC}$, then use the property of alternate interior angles. For polygons of more sides: cut them into triangles, and get the formula $(n-2)180°$. For sum of exterior angles: Pretend the polygon's sides are streets, you are walking along the streets. Each time you turn a corner, you turn an exterior angle; when you return to the starting point, you turned a total of $360°$. You can also use the formula to calculate it.

Problem 4.3 Prove the following: in a triangle, an exterior angle equals the sum of the two interior angles not adjacent to it.

Solution: The exterior angle is supplementary to the adjacent interior angle. Also the sum of all three interior angles is $180°$. The proof is obtained by comparing these two statements.

Problem 4.4 In a regular hexagon $ABCDEF$:

(a) Is ACE an equilateral triangle? Justify your answer.

Solution: Yes.
Note using SAS, $\triangle ABC \cong \triangle CDE \cong EFA$. Hence $AC = CE = EA$ so $\triangle ACE$ is an equilateral triangle.

(b) Calculate the angle AED.
Solution: 90°.
Note that $\triangle AFE$ is isosceles with $\angle AFE = 120°$. Thus, $\angle FEA = (180 - 120)/2 = 30°$ and $\angle AED = \angle FED - \angle FEA = 120 - 30 = 90°$.

Problem 4.5 Suppose that $ABCD$ is a square. Let point E be *outside* the square and that $\triangle CDE$ is an equilateral triangle (see the diagram). What is the measure of $\angle EAD$?

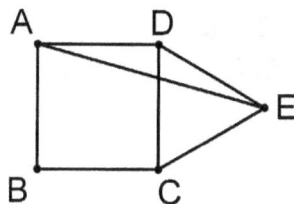

Solution: Since $DE = DC = DA$, $\triangle ADE$ is isosceles, therefore $\angle EAD = \angle DEA$. We know that $\angle ADE = 90° + 60° = 150°$, and $\angle ADE + \angle EAD + \angle DEA = 180°$, thus $\angle EAD = \dfrac{1}{2}(180° - 150°) = 15°$.

Problem 4.6 Same question as above, but with E *inside* the square.
Solution: 75°.
We have $\angle APD = 90 - 60 = 30°$. As $\triangle APD$ is isosceles we therefore have $\angle PAD = (180 - 30)/2 = 75°$.

Problem 4.7 Given square $ABCD$, let P and Q be the points outside the square that make triangles CDP and BCQ equilateral. Segments $\overline{AQ}$ and $\overline{BP}$ intersect at G. Find angle AGP.
Solution: 90°.
$\angle CBP = \angle BAQ = 15°$ based on previous problems. So $\angle GBA = 90 - 15 = 75°$, and then $\angle AGP = 180 - 75 - 15 = 90°$.

Problem 4.8 Mark P inside square $ABCD$, so that triangle ABP is equilateral. Let Q be the intersection of BP with diagonal AC. Triangle CPQ looks isosceles. Is this actually true?

Solution: True.
Based on previous problems, $\angle BPC = \angle BCP = 75°$. Since $\angle BCA = 45°$, we get $\angle PCQ = 75° - 45° = 30°$. Hence $\angle PQC = 75° = \angle CPQ$. So $\triangle CPQ$ is a $30°$-$75°$-$75°$ triangle.

Problem 4.9 Point P is inside regular pentagon $ABCDE$ so that triangle ABP is equilateral. Decide whether or not quadrilateral $ABCP$ is a parallelogram, and give your reasons.
Solution: No.
We have $\angle PAB = 60°, \angle ABC = 108°$ from the fact we have an equilateral triangle and a regular pentagon. Hence $ABCP$ is not a parallelogram as the angles are not supplementary. Further, note $\triangle BCP$ is isosceles, with $\angle PBC = 108 - 60 = 48°$, so $\angle BCP = \angle CPB = (180 - 48)/2 = 66°$. This means $\angle CPA = 66 + 60 = 126°$, so the quadrilateral $ABCP$ has angles $60°, 108°, 66°, 126°$, not a parallelogram.

Problem 4.10 In $\triangle ABC$, $\angle C = 2\angle A$, $AC = 2BC$. Find the measure of $\angle B$.
Solution: $90°$.
Construct the perpendicular bisector of $\overline{AC}$, intersecting $\overline{AB}$ at E, as in the diagram below

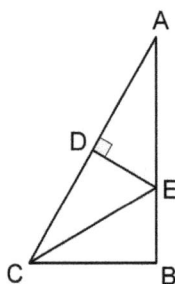

Then (using SAS) $\triangle ADE \cong \triangle CDE$. Since $\angle A = \angle DCE$ and $\angle C = 2\angle DCE$, we also have (using SAS) $\triangle EDC \cong \triangle EBC$. Hence $\angle B = \angle EDC = 90°$.

Problem 4.11 In $\triangle ABC$, $\angle ABC = 12°$, $\angle ACB = 132°$. Let $\overline{BM}$ and $\overline{CN}$ be the exterior angle bisectors where M and N are on the lines $\overline{AC}$ and $\overline{AB}$ respectively. Then
(A) $BM > CN$ (B) $BM = CN$ (C) $BM < CN$ (D) Can't determine.
Solution: (B)
It will help if an accurate diagram is drawn.

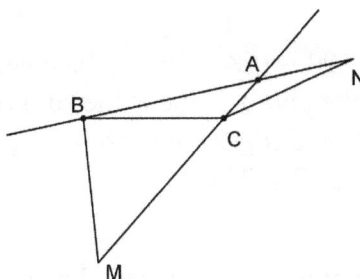

Since $\overline{BM}$ is the exterior angle bisector of $\angle ABC$, $\angle MBC = (180° - 12°)/2 = 84°$. Also, $\angle MCB = 180° - 132° = 48°$, hence $\angle BMC = 180° - 84° - 48° = 48°$. So $BM = BC$. Since $\overline{CN}$ is the exterior angle bisector of $\angle ACB$, $\angle NCA = (180° - 132°)/2 = 24°$. Also, $\angle BNC = 180° - \angle ABC - (\angle ACB + \angle NCA) = 180° - 12° - (132° + 24°) = 12° = \angle ABC$, thus $BC = CN$.
Therefore, $BM = CN$.

Problem 4.12 Given square $ABCD$, let P and Q be the points outside the square that make triangles CDP and BCQ equilateral. Prove that triangle APQ is also equilateral. What if $ABCD$ is a rectangle? What if $ABCD$ is a parallelogram?
Solution: APQ is equilateral in all the cases.
To prove $AP = PQ = QA$, we prove $\triangle ADP \cong \triangle QBA \cong \triangle QCP$ via SAS congruency. We always have $AD = BC = BQ = CQ$ and $AB = CD = CP = PD$ because we have a parallelogram and equilateral triangles. In the cases of square and rectangle, $\angle ADP = \angle QBA = \angle QCP = 90 + 60 = 150°$. In the case of general parallelogram, $\angle ADP = \angle QBA = 60° + \angle ABC = 360 - 60 - 60 - (180 - \angle ABC) = \angle QCP$ as needed.

Problem 4.13 Equilateral triangles BCP and CDQ are attached to the outside of regular pentagon $ABCDE$. Is quadrilateral $BPQD$ a parallelogram? Justify your answer.
Solution: No.
We have $\triangle PCQ$ is isosceles, with $\angle PCQ = 360 - 60 - 60 - 108 = 132$, hence $\angle CPQ = \angle CQP = (180 - 132)/2 = 24°$. Hence $\angle PQD = \angle QPB = 24 + 60 = 84°$ and hence $BPQD$ is not a parallelogram. In fact, similar reasoning ($\triangle BCD$ is isosceles) allows us to calculate the remaining angles as $96°$, so in fact $BPQD$ is a trapezoid.

Problem 4.14 Three non-overlapping regular plane polygons all have sides of length 1. The polygons meet at a point A in such a way that the sum of the three interior angles at A is $360°$. Thus the three polygons form a new polygon P (not necessarily convex) with A as an interior point. Among the three polygons, one is a square and one is a pentagon. Find the perimeter of P.

Solution: 23.

The remaining angle is $360 - 90 - 108 = 162°$, which belongs to an icosagon (20 sides). Therefore the perimeter of the entire shape is the sum of the perimeter of each of the three polygons minus the 3 shared edges. Note each shared edge is counted twice. Hence the perimeter is $4 + 5 + 20 - 3 \cdot 2 = 23$.

Problem 4.15 Find the side of the largest square that can be drawn inside an equilateral triangle with side length 12, one side of the square aligned with one side of the triangle.

Solution: $24\sqrt{3} - 36$.

Let x be the side length. Note that the square divides the triangle into two congruent 30-60-90 triangles and a smaller equilateral triangle (with side length x). Since the altitude of an equilateral triangle with side length s is $s\sqrt{3}/2$ we have: $\frac{\sqrt{3}}{2}x + x = 12 \cdot \frac{\sqrt{3}}{2}$. Solving for x gives $x = 24\sqrt{3} - 36$.

Problem 4.16 The equiangular convex hexagon $ABCDEF$ has $AB = 1$, $BC = 4$, $CD = 2$, and $DE = 4$. Find $[ABCDEF]$.

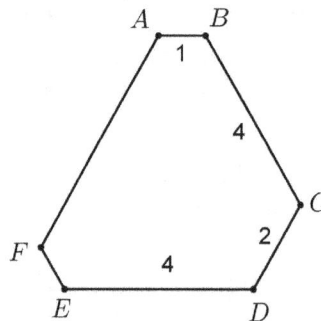

Solution: $43\sqrt{3}/4$.

Extend the sides $\overline{AF}, \overline{BC}, \overline{DE}$ to both ends to form a large equilateral triangle of side length 7, as shown.

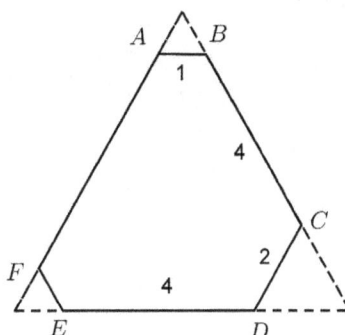

Then $[ABCDEF]$ is the area of the large equilateral triangle minus three smaller equilateral triangles of sides 1, 1, and 2. As the area of an equilateral triangle with side length s is $s^2\sqrt{3}/4$, we have $[ABCDEF] = (7^2 - 2^2 - 1^2 - 1^2) \times \sqrt{3}/4 = 43\sqrt{3}/4$.

Problem 4.17 A stop sign — a regular *octagon* — can be formed from a 12-inch square sheet of metal by making four straight cuts that snip off the corners. How long are the sides of the resulting polygon?

Solution: $12(\sqrt{2} - 1)$.

Let x be the side length of the octagon. The four cuts at the corners form 4 congruent 45-45-90 triangles with hypotenuse x. Hence the other two sides of each of these triangles is $x/\sqrt{2}$. As the original side of the square consists of the side of the octagon and two triangles, we have $x + 2 \cdot \dfrac{x}{\sqrt{2}} = 12$. Solving for x we get $x = 12(\sqrt{2} - 1)$.

Problem 4.18 The Golden Triangle. In $\triangle ABC$, $AB = AC$. Point D is on side $\overline{AB}$ such that $\overline{CD}$ bisects $\angle ACB$, and $CD = BC$. Find the angles of $\triangle ABC$.

Solution: $\angle A = 36°, \angle B = \angle C = 72°$.

Consider the diagram below.

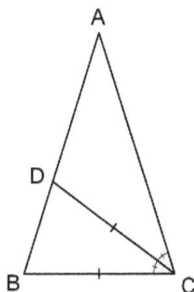

$AB = AC$ so $\angle B = \angle C$. From $CD = BC$ we get $\angle B = \angle BDC$. So $\angle A = 180° - \angle B - \angle C = 180° - \angle B - \angle BDC = \angle DCB = \dfrac{1}{2}\angle C$. So $180° = \angle A + \angle B + \angle C = 5\angle A$, thus

$\angle A = 36°$ and $\angle B = \angle C = 72°$.

Problem 4.19 $BCDE$ is a square, $\triangle ABC \cong \triangle FCD$ with $\angle A = 120°$ and $AB = AC$. If $AF = 20$, compute the area of square $BCDE$.

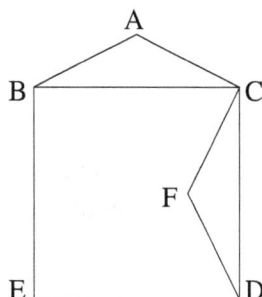

Solution: 600.
Since $\triangle ABC \cong \triangle FCD$ we have $AC = CF$ and $\angle ACF = 90°$. Hence $\triangle ACF$ is a 45-45-90 triangle, and thus $AC = AF/\sqrt{2} = 20/\sqrt{2} = 10\sqrt{2}$. Since $\angle A = 120°$ and $\triangle ABC$ is an isosceles triangle, $\angle ABC = \angle ACB = (180 - 120)/2 = 30°$. Hence, if M is the midpoint of $\overline{BC}$, then $\triangle AMC$ is a 30-60-90 triangle. Therefore, $MC = AC/2 = 5\sqrt{2}$ so $BC = 10\sqrt{2}$. This implies $[BCDE] = BC^2 = 600$.

Problem 4.20 In $\triangle ABC$, $\angle B = 90°$ and $\angle C = 30°$. Point D is on $\overline{BC}$ such that $\angle ADB = 45°$, and $DC = 10$. What is AB?
Solution: $5(\sqrt{3}+1)$.
We have $\triangle ABC$ is a 30-60-90 triangle and $\triangle ADB$ is a 45-45-90 triangle. Thus, if $x = AB$, then $BD = x$ as well. Also $BC = x\sqrt{3}$, so $CD = BC - BD = x(\sqrt{3} - 1) = 10$, thus $x = \dfrac{10}{\sqrt{3} - 1} = 5(\sqrt{3}+1)$.

Problem 4.21 Let $ABCD$ be a square. Through A construct line $\overline{AP}$ outside the square so that $\angle DAP = 45°$. Let E be a point on $\overline{AP}$ so that $BE = BD$. Let F be the intersection of $\overline{BE}$ and $\overline{AD}$. Is $\triangle DEF$ isosceles? Justify your answer.
Solution: Yes.
Assume the side length of the square is 1. Draw $\overline{EG} \perp \overline{BD}$ at G resulting in the diagram below.

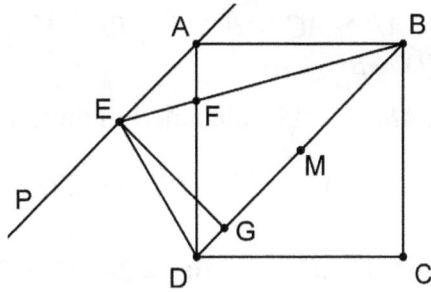

Since $\overline{AP}$ is parallel to $\overline{BD}$, we have that $EG = AP = \sqrt{2}/2$. Since $BE = BD = \sqrt{2}$ (combined with $\angle BGE = 90°$) we have $\triangle BEG$ is a 30-60-90 triangle. Hence, as $\triangle BED$ is isosceles, $\angle BED = \angle BDE = (180 - 30)/2 = 75°$. Further, $\angle ABF = 45 - 30 = 15°$, so $\angle EFD = \angle AFB = 180 - 90 - 15 = 75°$. Therefore $\triangle DEF$ is isosceles as claimed.

Problem 4.22 Let ABC be an isosceles right triangle with $\angle C = 90°$. Let P be a point inside the triangle such that $AP = 3, BP = 5$, and $CP = 2\sqrt{2}$. What is $[ABC]$?

Solution: Rotate $\triangle BCP$ 90° about C so that point B ends at A. In other words, let P' be the point outside $\triangle ABC$ so that $AP' = BP = 5$ and $CP' = CP = 2\sqrt{2}$, as in the diagram below.

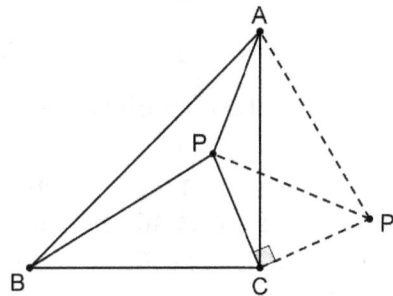

So $\triangle PCP'$ is an isosceles right triangle, hence a 45-45-90 triangle and thus $PP' = 2\sqrt{2} \cdot \sqrt{2} = 4$. This implies that $\triangle APP'$ is a 3-4-5 triangle, which means $\angle APP' = 90°$. Thus $\angle APC = 90 + 45 = 135°$. Using Law of Cosines on $\triangle APC$, $AC^2 = 3^2 + (2\sqrt{2})^2 - 2 \cdot 3 \cdot 2\sqrt{2} \cos 135° = 29$. Therefore $[ABC] = AC^2/2 = 29/2$.

Problem 4.23 Prove the Pythagorean Theorem using areas.

Solution: There are literally hundreds of different proofs. The simplest: Let $\triangle ABC$ be a right triangle where $\angle C$ is a right angle. Label the sides with a, b, c in the standard way. Draw the altitude $\overline{AD}$ on the hypotenuse. Then the three triangles, $\triangle ABC, \triangle ADC$ and $\triangle BDC$ are all similar. Their side ratio is $a : b : c$ based on the hypotenuses. Thus their area ratio is $a^2 : b^2 : c^2$. Since the two smaller triangle's areas add up to that of the big triangle, we have $a^2 + b^2 = c^2$.

Problem 4.24 In $\triangle ABC$, $AB > AC > BC$, $\overline{CD}, \overline{BE}, \overline{AF}$ are altitudes on $\overline{AB}, \overline{AC}, \overline{BC}$, respectively. Show that $CD < BE < AF$.

Solution: The area of $\triangle ABC$ can be calculated in three ways: $[ABC] = \frac{1}{2}AB \cdot CD = \frac{1}{2}AC \cdot BE = \frac{1}{2}BC \cdot AF$. Hence if $AB > AC > BC$ then $CD < BE < AF$.

Problem 4.25 In triangle ABC, $AC = 10$, $BC = 24$, $AB = 26$. What is the altitude on $\overline{AB}$?
Solution: $\frac{120}{13}$.
Note $\triangle ABC$ is a right triangle. Therefore the area is $10 \cdot 24/2 = 120$. Since $AB = 26$ the area is also $26h/2 = 13h$ where h is the altitude on $\overline{AB}$, we have $h = 120/13$.

Problem 4.26 Let $ABCD$ be a parallelogram, and E, H, F, G be points on sides $\overline{AB}, \overline{BC}$, $\overline{CD}, \overline{DA}$ respectively, and $\overline{EF} \| \overline{BC}$ and $\overline{GH} \| \overline{AB}$. Let P be the intersection of $\overline{EF}$ and $\overline{GH}$. If $[GPFD] = 10$, $[PHCF] = 8$, $[EBHP] = 16$, find $[ABCD]$.
Solution: 54.
Note $\overline{EF}, \overline{GH}$ divide $ABCD$ into four parallelograms. We have $[AEPG]/[GPFD] = [EBHP]/[PHCF]$ as $AEPG, GPFD$ and $EBHP, PHCF$ each share heights, and $AG = BH, GD = HC$. Thus $[AEPG] = 10 \cdot 16/8 = 20$. Hence the total area of $ABCD$ is $10 + 8 + 16 + 20 = 54$.

Problem 4.27 In $\triangle ABC$, let D, E, F be midpoints of the sides $\overline{BC}, \overline{AC}, \overline{AB}$. Show that $[DEF] = [ABC]/4$.
Solution: Note $\triangle ECD \triangle ACB$ (SAS) with ratio of corresponding sides $1:2$ so $DE = AB/2$. Similarly we have $EF = BC/2, EF = AC/2$. Hence $\triangle DEF \sim \triangle ABC$ (SSS) with ratio of sides $1:2$. Hence we have $[DEF]:[ABC] = 1:4$ as needed.

Problem 4.28 Suppose you only know that the centroid exists. Prove (using areas!) that the centroid divides each median in a ratio of $1:2$.
Solution: Consider the following diagram with the medians intersecting at G:

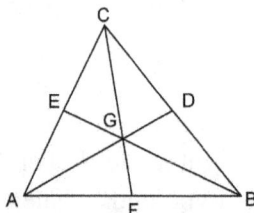

Note first that $[ACF] = [BCF]$ and $[AGF] = [BGF]$ (F is a midpoint and they share respective heights). Hence $[ACG] = [ACF] - [AGF] = [BCF] - [BGF] = [BCG]$. An identical argument gives $[ACG] = [ABG]$. Hence, $[AFG] = [ABG]/2 = [ACG]/2$ so in particular $[AFG] = [AFC]/3$. These two triangles share a height from A, so their bases FG, FC must be in ratio $1:3$ and hence $FG:GC = 1:2$. The other medians are handled identically.

Problem 4.29 Let $ABCD$ be a parallelogram with $[ABCD] = 1$. Let P be a point in the interior of $ABCD$. Show that $[ABP] + [CDP]$ is a fixed value and find that value.
Solution: $1/2$.
Note the height from $\overline{AB}$ to $\overline{CD}$ through P splits into altitudes for the two given triangles. As these altitudes sum to the total height of the parallelogram, $[ABP] + [CDP] = [ABCD]/2$.

Problem 4.30 Let $ABCD$ be a parallelogram, with midpoints E, F, G, H (say on $\overline{AB}$, $\overline{BC}$, $\overline{CD}, \overline{DA}$). Let I, J be the midpoints of $\overline{EF}, \overline{GH}$. Find the area of $\triangle JIG$ as a fraction of the area of $ABCD$.
Solution: $1/8$.
Using congruence we can prove that $\overline{EF} \| \overline{GH}$ and $\overline{FH} \| \overline{DC}$. We then have $[JIG] = [HIG]/2$ (as both triangle share the same height and J midpoint of $\overline{GH}$). Further $[HIG] = [HFG]$ (as $\overline{EF} \| \overline{GH}$ so both triangles have the same height and same base) and then $[HFG] = [DCFH]/2$ (as the triangle and parallelogram share the same height and base), and finally $[DCFH] = [ABCD]/2$ (since F, H are midpoints and both are parallelograms). Combining all of the above gives $[JIG] = [ABCD]/8$.

Problem 4.31 Prove that if a triangle has side lengths a, b, c, inradius r, and circumradius R we have $2Rr = \dfrac{abc}{a+b+c}$.
Solution: We have that the area of the triangle is $abc/4R$ and sr where s is the semiperimeter, so setting the equations equal we get that $abc = 4srR$. Rearranging and recalling $s = (a+b+c)/2$ gives the desired result.

Problem 4.32 Suppose the altitudes of a triangle are in ratio $2:2:3$ and the triangle has a perimeter of 24. Find the area of the triangle.
Solution: $18\sqrt{2}$.
As in an earlier problem, we use the fact that we can calculate the area using all of the altitudes. Suppose the altitudes (from A, B, C respectively) are h_A, h_B, h_C with $h_A : h_B : h_C = 2:2:3$. If a, b, c are the opposite sides, we then have (after cancelling $1/2$): $a \cdot h_A = b \cdot h_B = c \cdot h_C$. We therefore have that $b:a = h_A:h_B = 2:2$ and $c:b = h_B:h_C = 3:2$. Hence, the altitudes are in ratio $h_C:h_B:h_A = 2:3:3$. If the perimeter of the triangle is 24, this means the sides of the triangles are $6, 9, 9$. Using Heron's formula, the area is thus, $18\sqrt{2}$.

Problem 4.33 Given square $ABCD$, let E, F be the midpoints of AB and BC respectively, and G be the intersection of AF and CE. If $[ABCD] = 1$, find $[AGCD]$.

Solution: $2/3$.

Draw line segment $\overline{AC}$ dividing $ABCD$ into two triangles with area $1/2$. Note that $\overline{CE}, \overline{AF}$ are thus medians of $\triangle ABC$. Hence, using an earlier problem, $[ACG] = [ABC]/3 = 1/6$. Hence, $[AGCD] = [ABC] + [ACG] = 1/2 + 1/6 = 2/3$.

Problem 4.34 Suppose you have a trapezoid $ABCD$ with $\overline{AB}$ parallel to $\overline{CD}$. Let E be the intersection of the diagonals. Suppose $AB = 10, CD = 15$ and $\triangle ADE$ has area 24. Find the area of $ABCD$.

Solution: 100.

Since $\overline{AB} \| \overline{CD}$, $\triangle ABE \sim \triangle CDE$, with ratio of corresponding sides $10 : 15 = 2 : 3$. Hence $DE : EB = 3 : 2$ and since $\triangle AED, \triangle AEB$ share the same height from A, $[AED] : [AEB] = 3 : 2$ so $[AEB] = 24 \cdot \dfrac{2}{3} = 16$. We can use a similar argument to get $[DEC] = [ADE] \cdot \dfrac{3}{2} = 36$ and $[BEC] = [AEB] \cdot \dfrac{3}{2} = 24$. Hence the total area is $24 + 16 + 36 + 24 = 100$.

Problem 4.35 Prove the converse of the Pythagorean Theorem. Hint: You can use the Pythagorean Theorem!

Solution: Suppose we have triangle $\triangle ABC$ with $AB = c, AC = b, BC = a$, if $c^2 = a^2 + b^2$ then $\triangle ABC$ is a right triangle. Now form a right triangle $\triangle A'B'C'$ with $A'C' = b$ and $B'C' = a$. By the Pythagorean Theorem, $A'B' = c^2$. Hence, using SSS, $\triangle ABC \cong \triangle A'B'C'$ and therefore $\angle C = \angle C'$ is right.

Problem 4.36 Suppose you have a circle with diameter $\overline{AB}$ with $AB = 4$. Let C, D be on arc $\overset{\frown}{AB}$ such that $\overset{\frown}{AC} : \overset{\frown}{CD} : \overset{\frown}{DB} = 1 : 2 : 1$. Find the area of the figure enclosed by line segment $\overline{AC}$, arc $\overset{\frown}{CD}$, and line segment $\overline{AD}$.

Solution: π.

Let O be the center of the circle. Since $\overset{\frown}{AC} = \overset{\frown}{DB}$ we have $\triangle AOC \cong \triangle BOC$ so $\overline{CD} \| \overline{AB}$. Therefore $[ACD] = [OCD]$ where O is the center of the circle (as they share the same base and height). Hence the area of the figure we want is the same as the area of the sector with arc $\overset{\frown}{CD}$. Since $\overset{\frown}{CD}$ is half of arc $\overset{\frown}{AB}$ it is a quarter of a circle. Hence the area we want is $\dfrac{1}{4}\pi 2^2 = \pi$.

Problem 4.37 In triangle ABC, $AC = 9$, $BC = 10$, $AB = 17$. What is the altitude on $\overline{BC}$?

Solution: $36/5$.

This is not a right triangle, so we have to use Heron's Formula to calculate $[ABC] =$

$\sqrt{19\cdot(19-9)\cdot(19-10)\cdot(19-17)}=36$. Hence $36=BC\cdot h/2$ where h is the altitude from base $\overline{BC}$. We solve for h to get $h=36/5$.

Problem 4.38 Let $ABCD$ be a rectangle, and E,H,F,G be points on sides $\overline{AB}$, $\overline{BC}$, $\overline{CD}$, $\overline{DA}$ respectively, and $\overline{EF}\|\overline{BC}$ and $\overline{GH}\|\overline{AB}$. Let P be the intersection of EF and GH. Suppose $[GPFD]=10,[EBHP]=12$ and $[ABCD]=44$, with each of the four smaller rectangles having integer dimensions. Find all possible dimensions for $GPFD$ and $EBHP$.

Solution: $2\times5, 2\times6$ or $1\times10, 1\times12$.
Since we have integer coordinates, $GPFD$ is either 1×10 or 2×5. Similarly, $EBHP$ is either $1\times12, 2\times6$, or 3×4. The full rectangle can have dimensions $1\times44, 2\times22$, or 4×11. Trying out possibilities, we see only $2\times5, 2\times6$ or $1\times10, 1\times12$ work for the dimensions of $GPFD, EBHP$ respectively.

Problem 4.39 Find a formula for the area of a parallelogram if you are given the two sides lengths as well as one of the diagonals.
Solution: If the parallelogram has sides a,b with diagonal d, note the diagonal divides the parallelogram into two congruent triangles with sides a,b,d. Thus, if $s=(a+b+d)/2$, using Heron's formula we have the area of the parallelogram is $2\sqrt{s(s-a)(s-b)(s-d)}$.

Problem 4.40 Prove that if a triangle has side lengths a,b,c semiperimeter s, inradius r, and circumradius R we have

$$\frac{R}{r}=\frac{abc}{4(s-a)(s-b)(s-c)}.$$

Solution: Let A denote the area of the triangle. Rearranging area formulas we have $R=\dfrac{abc}{4A}, r=\dfrac{A}{s}$ so $\frac{R}{r}=\frac{sabc}{4A^2}$. Using Heron's we have $A^2=s(s-a)(s-b)(s-c)$ which completes the proof after simplifying.

Problem 4.41 Suppose a parallelogram P_1 has area 256. Connect the midpoints of each side to form a parallelogram P_2. Repeat to get P_3, and continue repeating until you get to P_{10}.

(a) Prove this problem actually makes sense. That is, prove that if you connect the midpoints of a parallelogram you get another parallelogram.
Solution: Consider the diagram below.

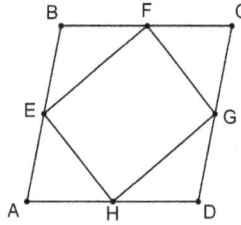

Note that since E,F,G,H are midpoints, $AH = CF, AE = CG$. Further, since $ABCD$ is a parallelogram, $\angle A = \angle C$. Hence $\triangle EHA \cong GFC$. Combined with the fact that $\overline{AB}\|\overline{CD}, \overline{BC}\|\overline{AD}$ we get that $\overline{EF}\|\overline{GH}, \overline{FG}\|\overline{EH}$ so $EFGH$ is a parallelogram as needed.

(b) Find the area of P_{10}.
Solution: $1/2$.
Consider the diagram as in (a). Since E,F,G,H are all midpoints, we have that $[AEH] = [BFE] = [CGF] = [DHG] = [ABCD]/8$. Hence, $[EFGH] = [ABCD]/2$. In general, this implies $[P_{n+1}] = [P_n]/2$. Hence, P_{10} has area $1/2^9$ that of P_1, so $[P_{10}] = 256/2^9 = 1/2$.

Problem 4.42 Let $ABCD$ be a trapezoid with $\overline{AB}\|\overline{CD}$. Let E be the intersection of the two diagonals $\overline{AC}, \overline{BD}$.

(a) If $ABCD$ is a parallelogram show $[ABE] = [BCE] = [CDE] = [DAE]$.
Solution: Recall the diagonals bisect each other in a parallelogram. Using diagonal $\overline{AC}$ we have that $[ABE] = [BCE], [DAE] = [CDE]$. Using diagonal $\overline{BD}$ we have $[BCE] = [CDE]$ which completes the proof.

(b) Prove that for a general trapezoid, $[BCE] = [DAE]$.
Solution: Since $\overline{AB} \| \overline{CD}$ we have $[ABC] = [ABD]$. The result follows as $[ABC] = [ABE] + [BCE], [ABD] = [ABE] + [DAE]$.

Problem 4.43 Let $\overline{AM}$ be a median of $\triangle ABC$, and D be a point on $\overline{MC}$, and E be a point on $\overline{AB}$, such that $\overline{ME} \| \overline{AD}$. Show that $[BDE] = [AEDC]$.
Solution: Let O be the intersection of $\overline{AM}$ and $\overline{DE}$, as in the diagram below.

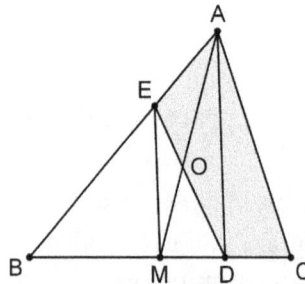

First note $[ABM] = [ACM]$ because M is the midpoint of $\overline{BC}$. Since $\overline{ME}\|\overline{AD}$, $AEDM$ is a trapezoid, so by an earlier problem, $[AOE] = [MOD]$. Hence, $[BEOM] = [ACOD]$ as $[BEOM] + [AEO] = [ABM] = [ACM] = [ACOD] + [MOD]$. Thus, $[BDE] = [BEOM] + [MOD] = [ACOD] + [AOE] = [AEDC]$ as needed.

Problem 4.44 Prove Ceva's Theorem using areas.
Solution: Let G be the intersecting point of the three cevians in the diagram below.

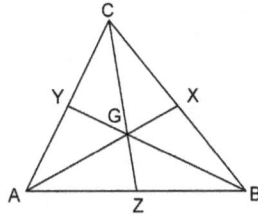

Then, as triangles $\triangle AXB, \triangle AXC$ and triangles $\triangle GXB, \triangle GXC$ both share the same height,

$$\frac{BX}{XC} = \frac{[ABX]}{[AXC]} = \frac{[GBX]}{[GXC]} = \frac{[ABX] - [GBX]}{[AXC] - [GXC]} = \frac{[ABG]}{[ACG]},$$

Similarly $\dfrac{CY}{YA} = \dfrac{[BCG]}{[BAG]}$, and $\dfrac{AZ}{ZB} = \dfrac{[CAG]}{[CBG]}$. Multiplying all three equations we get a product of 1, as needed.

Problem 4.45 Prove the converse to Ceva's Theorem using an indirect proof.
Solution: Let G be the intersection of AX and BY. Connect and extend CG to intersect with size AB at point Z'. Then the cevians AX, BY, and CZ' are concurrent. Based on Ceva's Theorem, $\dfrac{BX}{XC} \cdot \dfrac{CY}{YA} \cdot \dfrac{AZ'}{Z'B} = 1$. This means $\dfrac{AZ}{ZB} = \dfrac{AZ'}{Z'B}$. Thus Z and Z' are the same point. ∎

Problem 4.46 Prove the Angle Bisector Theorem using areas.
Solution: Draw perpendicular lines from D to sides $\overline{AB}$ and $\overline{AC}$, with feet E and F respectively, yielding the diagram below.

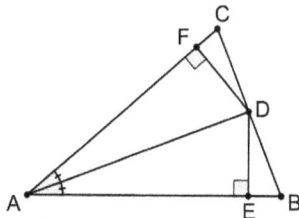

Since D is on the angle bisector of $\angle BAC$, $\triangle AED \cong \triangle AFD$ so the lengths $DE = DF$. The segment $\overline{AD}$ is a cevian, and the two triangles ABD and ACD have the same height from point A, thus the ratio between the areas $[ABD]/[ACD] = BD/DC$. On the other hand, $[ABD] = \frac{1}{2}AB \cdot DE$ and $[ACD] = \frac{1}{2}AC \cdot DF$, and with the fact $DE = DF$, we have

$$\frac{BD}{CD} = \frac{[ABD]}{[ACD]} = \frac{(1/2)AB \cdot DE}{(1/2)AC \cdot DF} = \frac{AB}{AC}.$$

Problem 4.47 Centers of Triangles: Prove the following, using Ceva's Theorem (or its converse) if applicable.

  (a) (Centroid). In every triangle ABC, the three medians (i.e. the line from a vertex to the midpoint of the opposite side) are concurrent. This point is called the *centroid* of the triangle ABC.

   Solution: Recall the general diagram for Ceva's Theorem.

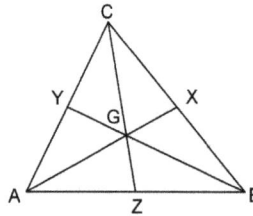

   Since X, Y, Z are midpoints, $\dfrac{BX}{XC} \cdot \dfrac{CY}{YA} \cdot \dfrac{AZ}{ZB} = 1 \cdot 1 \cdot 1 = 1$ so the medians are concurrent by the converse of Ceva's Theorem.

  (b) (Incenter). In every triangle ABC, the three angle bisectors (i.e. the line from a vertex bisecting the angle at that vertex) are concurrent. This point is called the *incenter* of the triangle ABC, because it is the center of the circle inscribed in triangle ABC.

   Solution: Consider the general diagram as in part (a). Since AX, BY, CZ are angle bisectors, the angle bisector theorem tells us that $\dfrac{BX}{XC} = \dfrac{AB}{AC}, \dfrac{CY}{YA} = \dfrac{BC}{AB}, \dfrac{AZ}{ZB} = \dfrac{AC}{BC}$ so $\dfrac{BX}{XC} \cdot \dfrac{CY}{YA} \cdot \dfrac{AZ}{ZB} = \dfrac{AB \cdot BC \cdot AC}{AC \cdot AB \cdot BC} = 1$. Hence the converse of Ceva's theorem says that the angle bisectors are concurrent.

  (c) (Orthocenter). In every triangle ABC, the three altitudes (i.e. the line from a vertex perpendicular to the opposite side) are concurrent. The common point of intersection is called the *orthocenter* of triangle ABC.

   Solution: Consider the general diagram as in part (a). Let the feet of the altitudes on BC, CA, AB be X, Y, Z respectively. Then $BX = AB\cos B$, $XC = AC\cos C$, etc. So $\dfrac{BX}{XC} \cdot \dfrac{CY}{YA} \cdot \dfrac{AZ}{ZB} = \dfrac{AB\cos B}{AC\cos C} \cdot \dfrac{BC\cos C}{AB\cos A} \cdot \dfrac{AC\cos A}{BC\cos B} = 1$. Thus by the converse to

Ceva's Theorem, the altitudes are concurrent.

(d) (Circumcenter). In every triangle ABC, the perpendicular bisectors of the three sides are concurrent. This point is called the *circumcenter* of the triangle ABC, because it is the center of the circle circumscribed around triangle ABC.

Solution: The property of perpendicular bisector is: any point on the perpendicular bisector is equidistant to the two endpoints of the line segment, and vice versa. Let the perpendicular bisectors of sides $\overline{AB}$ and $\overline{AC}$ intersect at O, then $OA = OB$ and $OA = OC$, so $OB = OC$, which means O is also on the perpendicular bisector of $\overline{BC}$. ∎

Second Solution: One way to use the Ceva's Theorem is the following: connect the midpoints D, E, F of the sides $\overline{BC}, \overline{CA}, \overline{AB}$ to form a triangle DEF. Then the three perpendicular bisectors of $\triangle ABC$ are in fact the three altitude of $\triangle DEF$ as in the diagram below.

We have proven the three altitude of any triangle are concurrent (the orthocenter of $\triangle DEF$) using Ceva's, thus the three perpendicular bisectors of $\triangle ABC$ are concurrent.

Problem 4.48 In $\triangle ABC$, $AB = AC$, and P is a point on the side $\overline{BC}$. Prove that the sum of the distances from P to the sides $\overline{AB}$ and $\overline{AC}$ is a fixed value.

Solution: Connect $\overline{AP}$, forming triangles $\triangle ABP, \triangle ACP$. Let h_1, h_2 denote the heights of these triangles (where the height is the distance from P to the sides $\overline{AB}$ and $\overline{AC}$ respectively). Therefore, $[ABC] = [ABP] + [ACP] = h_1 \cdot AB/2 + h_2 \cdot AC/2$. We also have that $[ABC] = h \cdot AC/2$ where h is the height from B to $\overline{AC}$. Hence, since $AB = AC$ we have $h_1 + h_2 = h$, a fixed value.

Problem 4.49 Let G be the centroid of $\triangle ABC$, and $AG = 3, BG = 4, CG = 5$, find $[ABC]$.

Solution: 18.
Say the medians are $\overline{AD}, \overline{BE}, \overline{CF}$, which divide $\triangle ABC$ into six equal area triangles. Further, as the centroid divides each median in a ratio of $2 : 1$, we have that $GD = 3/2, GE = 2, GF = 5/2$. Extend $\overline{BE}$ 2 units further, giving a point H. Note $AGCH$ is a parallelogram (it's diagonals bisect each other), so $CH = 3$. Hence, $\triangle GCH$ is a $3, 4, 5$-right triangle with area 6. Note that $[GCH] = [ACG] = [ABC]/3$ and thus $[ABC] = 18$.

Problem 4.50 Let ABC be a triangle with ω as its circumcircle. Arcs $\overset{\frown}{AB}$, $\overset{\frown}{BC}$, and $\overset{\frown}{CA}$ have lengths 3,4, and 5, respectively. Find the area of triangle ABC.

Solution: Let O be the center of the circumcircle. ω has circumference $3+4+5=12$, so the circumradius is $\dfrac{6}{\pi}$. Also, as the arcs $\overset{\frown}{AB}, \overset{\frown}{BC}, \overset{\frown}{CA}$ are in ratio $3,4,5$, we have $\angle AOB = 90°, \angle BOC = 120°, \angle COA = 150°$. $[ABC] = [ABO] + [BCO] + [CAO]$, where each of these triangles are isosceles triangles. Thus

$$[ABC] = \frac{1}{2}\left[\left(\frac{6}{\pi}\right)^2 \sin 90° + \left(\frac{6}{\pi}\right)^2 \sin 120° + \left(\frac{6}{\pi}\right)^2 \sin 150°\right] = \frac{27 + 9\sqrt{3}}{\pi^2}$$

as needed.

Problem 4.51 (2002 AMC 12A) Triangle ABC is a right triangle with $\angle ACB$ as its right angle, $m\angle ABC = 60°$, and $AB = 10$. Let P be randomly chosen inside $\triangle ABC$, and extend $\overline{BP}$ to meet $\overline{AC}$ at D. What is the probability that $BD > 5\sqrt{2}$?

Solution: First note triangle ABC is a 30-60-90 triangle, so $BC = 5$, $AC = 5\sqrt{3}$. Let E be the point on $\overline{AC}$ such that $CE = 5$, as in the diagram below.

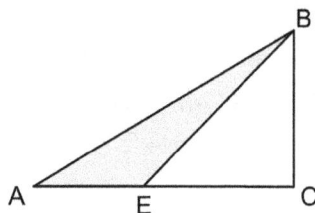

Then $BE = 5\sqrt{2}$ as $\triangle BCE$ is a 45-45-90 triangle. Hence, if P is inside $\triangle ABE$, then $BD > 5\sqrt{2}$. Therefore, the probability equals the ratio $\dfrac{[ABE]}{[ABC]} = \dfrac{AE}{AC} = \dfrac{5\sqrt{3} - 5}{5\sqrt{3}} = \dfrac{3 - \sqrt{3}}{3}$.

Problem 4.52 Let M be the intersection of line segments $\overline{AB}$ and $\overline{PQ}$. Show that $\dfrac{[PAB]}{[QAB]} = \dfrac{PM}{QM}$.

Solution: Since $\dfrac{[APM]}{[AQM]} = \dfrac{PM}{QM}$ (the two triangles have the same height), and $\dfrac{[BPM]}{[BQM]} = \dfrac{PM}{QM}$ (similar reason), the result is obtained (remember if $a : b = c : d$ then $a + c : b + d = a : b$).

Problem 4.53 In the parallelogram $ABCD$, E is a point on $\overline{AD}$, F is a point on $\overline{AB}$, and $BE = DF$. Also $\overline{BE}, \overline{DF}$ intersect at G. Show that $\angle BGC = \angle DGC$.

Solution: Connect $\overline{CE}$ and $\overline{CF}$. We have $[BCE] = [DCF] = \frac{1}{2}[ABCD]$, as both share a side and height with the parallelogram (it doesn't matter that they share different sides and heights). Construct heights $\overline{CH}, \overline{CI}$ for $\triangle BCE, \triangle DCF$ respectively (that is $\overline{CH} \perp \overline{BE}$ at H and $\overline{CI} \perp \overline{DF}$ at I). Since $[BCE] = [DCF]$ the heights are equal $CH = CI$. Thus C is on the angle bisector of $\angle BGD$. Therefore $\angle BGC = \angle DGC$.

Problem 4.54 ABC is a triangle with integer side lengths. Extend $\overline{AC}$ beyond C to point D such that $CD = 120$. Similarly, extend $\overline{CB}$ beyond B to point E such that $BE = 112$ and $\overline{BA}$ beyond A to point F such that $AF = 104$. If triangles CBD, BAE, and ACF all have the same area, what is the minimum possible area of triangle ABC?

Solution: Since $[CBD] = [BAE] = [ACF]$, we get $\dfrac{[CBD]}{[ABC]} = \dfrac{[BAE]}{[ABC]} = \dfrac{[ACF]}{[ABC]}$. Since $\triangle CBD, \triangle ABC$ share a same height, and the same is true for pairs $\triangle BAE, \triangle ABC$ and $\triangle ACF, \triangle ABC$, we can cancel these heights to get $\dfrac{120}{CA} = \dfrac{112}{BC} = \dfrac{104}{AB}$. Simplifying, $\dfrac{15}{CA} = \dfrac{14}{BC} = \dfrac{13}{AB}$. Since AB, BC, CA are all integers, the minimum possible values are $AB = 13, BC = 14, CA = 15$, so by Heron's formula, the minimum area is $[ABC] = \sqrt{21(21-13)(21-14)(21-15)} = \sqrt{(3 \cdot 7) \cdot (2^3) \cdot 7 \cdot (2 \cdot 3)} = 84$.

Problem 4.55 Given an equilateral triangle ABC and a point P in the interior of $\triangle ABC$. Show that the sum of the distances from P to all three sides is equal to the altitude of $\triangle ABC$.

Solution: Connect P and the three vertices. The areas of the three smaller triangles $\triangle PAB, \triangle PBC, \triangle PCA$ add up to $[ABC]$. If s, h are the side length and altitude respectively, we thus have $\frac{1}{2}sh = \frac{1}{2}(s \cdot PA + s \cdot PB + s \cdot PC)$ from which the result follows after simplifying.

Problem 4.56 Let $ABCD$ by a square with side length 1. Let E, P, F be the midpoints of AD, CE, BP respectively. Find $[BFD]$.

Solution: $1/16$.

Use the diagram below to help visualize the setup:

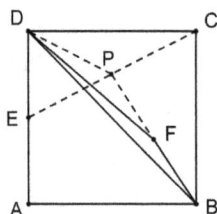

First note $[BFD] = [BPD]/2$ as they share the same height and F is a midpoint. Note $\frac{1}{2} = [BCD] = [BPC] + [DPC] + [BPD]$. Further $[BPC] = [ECB]/2 = 1/4$ as E, P are midpoints and $[DPC] = [DEC]/2 = 1/8$ as P is a midpoint. Hence $[BPD] = 1/8$ and $[BFD] = 1/16$.

Problem 4.57 Let P be an interior point in parallelogram $ABCD$, and $[APB] : [ABCD] = 2 : 5$. Find $[CPD] : [ABCD]$

Solution: $1 : 10$.

Since $ABCD$ is a parallelogram (hence the heights of $\triangle APB, \triangle CPD$ add up to the height of $ABCD$), we have $[APB] + [CPD] = [ABCD]/2$. From this it is clear that $[CPD] : [ABCD] = 1 : 10$.

Problem 4.58 Let $\odot O$ be a circle with radius 1. Let A be a point outside $\odot O$, and $OA = 2$. Let AB be tangent to $\odot O$ at B, and chord $BC \parallel OA$. Connect AC. Find the sum of the areas of $\triangle ABC$ and the region enclosed by BC and the minor arc $\overset{\frown}{BC}$.

Solution: $\pi/6$.

First note that since $\overline{BC} \parallel \overline{OA}$ we have $[ABC] = [OBC]$ so the area we are looking for is simply sector BOC. Since $\triangle ABO$ is a right triangle with $OB = 1, OA = 2$, it is a 30-60-90 triangle and hence $\angle BOA = 60°$. Therefore $\angle CBO = 60°$. As $\triangle BOC$ is isosceles, we thus have $\angle BOC = 60°$ as well. Hence the sector is $1/6$ of the unit circle, so has area $\pi/6$.

Problem 4.59 Let E be the midpoint of BC in parallelogram $ABCD$, G be the intersection of AE and BD. If $[BEG] = 1$, find $[ABCD]$.

Solution: 12.

Draw the other diagonal $\overline{AC}$ and let F be the intersection of $\overline{AC}, \overline{BD}$, so F is the midpoint of $\overline{AC}$. Hence $\overline{AE}, \overline{BF}$ are medians in $\triangle ABC$ so $[ABC] = 6 \cdot [BEG] = 6$. Therefore, $[ABCD] = 12$.

Problem 4.60 Let O be the intersection of the diagonals of convex quadrilateral $ABCD$. Given that $[ABC] = 5, [ACD] = 10$, and $[ABD] = 6$, find $[ABO]$.

Solution: From Problem 4.52 above, $\dfrac{BO}{OD} = \dfrac{[ABC]}{[ACD]} = \dfrac{1}{2}$, so $\dfrac{BO}{BD} = \dfrac{BO}{BO + OD} = \dfrac{1}{3}$.

Also, $\dfrac{[ABO]}{[ABD]} = \dfrac{BO}{BD} = \dfrac{1}{3}$, thus $[ABO] = 6 \cdot \dfrac{1}{3} = 2.$

Problem 4.61 (2002 AMC 12A) In triangle ABC, side $\overline{AC}$ and the perpendicular bisector of $\overline{BC}$ meet in point D, and $\overline{BD}$ bisects $\angle ABC$. If $AD = 9$ and $DC = 7$, what is the area of triangle ABD?

Solution: Consider the diagram below, where M is the midpoint of $\overline{BC}$.

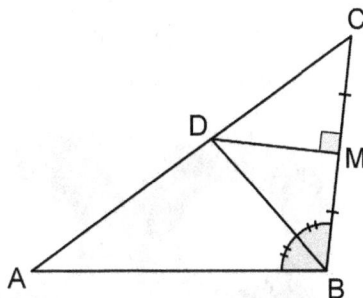

Using Hypotenuse-Leg, $\triangle CMD \cong \triangle BMD$, so $\angle DCM = \angle DBM$ and also $BD = CD = 7$. Since $\angle ABD = \angle DBC = \angle ACB$ and they share $\angle A$, we get $\triangle ABD \sim \triangle ACB$, so $\dfrac{AB}{AD} = \dfrac{AC}{AB}$, thus $\dfrac{AB}{9} = \dfrac{16}{AB}$, and then $AB = 12$. By Heron's formula, $[ABD] = \sqrt{14(14-7)(14-9)(14-12)} = 14\sqrt{5}.$

Problem 4.62 Let $\triangle ABC$ be a right triangle, where $\angle C$ is the right angle. Construct squares $ACDE$ and $BCFG$ on the outside of $\triangle ABC$. Assume that $\overline{AG}$ intersects $\overline{BC}$ at point P, and $\overline{BE}$ intersects $\overline{AC}$ at point Q, show that $CP = CQ$.

Solution: We show that $[APF] = [BQD]$. Connect $\overline{PF}$ and $\overline{CG}$ as in the diagram below.

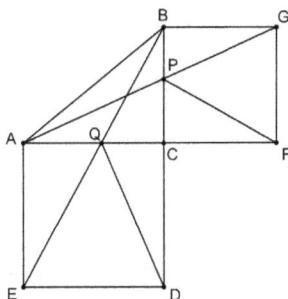

Therefore, as they share a base and height, $[CPG] = [CPF]$. Therefore, $[APF] = [APC] + [CPF] = [APC] + [CPG] = [ACG]$ Further, we have $[ACG] = [ABC]$ (same base and height) so in fact $[APF] = [ABC]$. Similarly, $[BQD] = [ABC]$ as well, so $[APF] = [BQD]$. Then, since the bases $AF = AC + CF = AC + BC = DC + CB = BD$ are equal, the altitudes $CP = CQ$ are equal as well.

Problem 4.63 Prove that in a circle, a radius is perpendicular to a chord if and only if the radius bisects the chord.

Solution: Let the chord be $\overline{AB}$ and say the radius is $\overline{OC}$ (where O is the center of the circle. Let the two intersect at E. Examine triangles $\triangle AOE, \triangle BOE$. If $\overline{OC} \perp \overline{AB}$, then $\triangle AOE \cong \triangle BOE$ (SAS) so $AE = EB$ as needed. If $AE = EB$, then $\triangle AOE \cong \triangle BOE$ (SSS) so $\angle AEO = \angle BEO$ and since the two angles are supplementary, both are right as needed.

Problem 4.64   (a) Suppose two tires, each with radius 1ft rest upright on the ground and touching each other, as pictured below:

How much space is needed horizontally to store the tires?
Solution: 4.
Note the distance between the center of the circles is parallel to the ground, so the total distance needed is the two diameters, which sum to 4ft.

(b) Repeat part (a) with two tires of radii 1 and 2 feet respectively.
Solution: $3 + 2\sqrt{2}$.
To store the tires using the least amount of horizontal space, we again examine the line $\overline{AB}$ connecting the two centers, as in the diagram below:

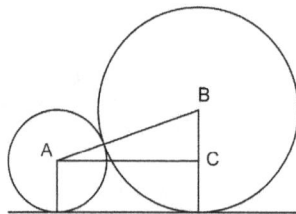

From here we see the horizontal distance in between the centers of the tires is AC, where $\triangle ABC$ is a right triangle with $AB = 1 + 2 = 3, BC = 2 - 1 = 1$. Hence $AC = \sqrt{3^2 - 1^2} = 2\sqrt{2}$ and the total distance needed is $1 + 2\sqrt{2} + 2 = 3 + 2\sqrt{2}$.

Problem 4.65 Suppose you start with a circle of radius 1. Pick a point on the circle and draw another circle of radius 1 (having the point as its center). Let R denote the region consisting of all points that are inside both circles.

(a) Find the perimeter of R.

Solution: $\dfrac{4\pi}{3}$.

Let A, B be the centers of the circle, and C, D be the intersection points of the two circles. Hence the perimeter of R consists of the arc from C to D (containing A) plus the arc from D to C (containing B). Further, as both $\triangle ABC, \triangle ABD$ are equilateral triangles (SSS), each arc is $120°$, or $1/3$ of a circle. Hence the perimeter of R is $2 \cdot \dfrac{1}{3} \cdot 2\pi = \dfrac{4\pi}{3}$.

(b) Find the area of R.

Solution: $\dfrac{2\pi}{3} - \dfrac{\sqrt{3}}{2}$.

Consider the labeling from above. We then have the area of R is the sum of the areas of the two sectors minus the two equilateral triangles: $2 \cdot \dfrac{1}{3} \cdot \pi - 2 \cdot \dfrac{\sqrt{3}}{4}$.

Problem 4.66   (a) What is the radius of the largest circle that can fit in a quarter circle of radius 1?

Solution: $\sqrt{2} - 1$.

Consider the following general diagram:

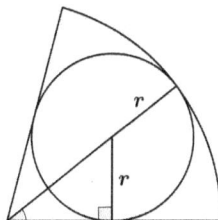

where if the sector is $\theta°$, the marked angle in the right triangle is $\theta/2$.

In this case, $\theta = 90°$, so the triangle is a 45-45-90 triangle with hypotenuse $r\sqrt{2}$.

Hence, $1 = r + r\sqrt{2}$ so solving for r gives $r = \dfrac{1}{1+\sqrt{2}} = \sqrt{2} - 1$.

(b) What if instead you are fitting it into a $60°$-sector?

Solution: $1/3$.

A similar method to part (a) (now we have a 30-60-90 triangle so the hypotenuse is $2r$) gives the equation $1 = r + 2r$ so $r = 1/3$.

Problem 4.67 Arrange 4 congruent circles so that (i) the center of the four circles form a square with side length 10, and (ii) adjacent circles are tangent.

(a) Find the radius of each circle.

Solution: 5.

From the given information we have that the radius of each circle is half the side length, so $10/2 = 5$.

(b) Find the area of the region inside the square that is outside each of the circles.
Solution: $100 - 25\pi$.
The total square has area $10^2 = 100$. Note a quarter of each circle is inside the square, so these quarter circles can be combined to create a full circle of radius 5. Hence the area we want is the area of the square minus the area of the circle: $100 - \pi 5^2$.

Problem 4.68 Prove that in a circle, the perpendicular bisector of a chord passes through the center of the circle.
Solution: Label the chord $\overline{AB}$ with midpoint C. Draw a radius of the circle that passes through C. Since this radius bisects the chord, it is perpendicular to the chord, hence the perpendicular bisector passes through the center as needed.

Problem 4.69 Suppose you have three tires with radii $1, 2, 3$ feet respectively. You store them horizontally as in Problem 4.64 above. What is the minimum amount of horizontal space needed to store the tires? Justify your answer!
Solution: $5 + 2\sqrt{3} + 2\sqrt{2}$.
Note reversing the order of the tires does not change the number of horizontal space needed, so we need to examine three arrangements of the tires, as in the diagram below:

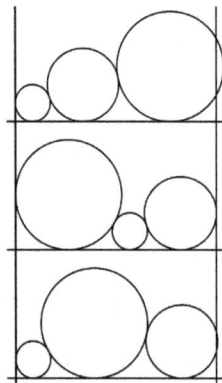

You can then proceed similar to the 4.64 part (b) above, to calculate the horizontal lengths needed for each arrangement to be respectively $1 + \sqrt{8} + \sqrt{24} = 4 + 2\sqrt{2} + 2\sqrt{6}, 3 + \sqrt{12} + \sqrt{8} + 2 = 5 + 2\sqrt{3} + 2\sqrt{2}, 1 + \sqrt{12} + \sqrt{24} + 2 = 3 + 2\sqrt{3} + 2\sqrt{6}$. A quick calculation shows the second arrangement is slightly smaller than the other two.

Problem 4.70 Suppose you start with a circle of radius 1. Pick a point on the circle and draw another circle of radius 1 (having the point as its center). Now pick one of the two intersection points and draw a third circle centered at this point, also with radius 1 (and again having the point as its center). Call R the region consisting of all points that are contained inside at least two circles.

  (a) Find the perimeter of R.

   Solution: 2π.

   The points of intersection of the circles form 4 equilateral triangles (with side length 1. Hence, the perimeter of R consists of six $60°$ arcs, hence is the same as the circumference of a unit circle: 2π.

  (b) Find the area of R.

   Solution: $\pi - \sqrt{3}/2$.

   As mentioned above, R contains 4 equilateral triangles with side length 1, having total area $4 \cdot \sqrt{3}/4 = \sqrt{3}$. The leftovers consist of 6 congruent circle segments. The area of each of these is a $60°$ sector minus an equilateral triangle with side length one: $\dfrac{1}{6}\pi - \dfrac{\sqrt{3}}{4}$, so in total they have area $\pi - 3\sqrt{3}/2$. Hence the total area of R is $\pi - \sqrt{3}/2$.

Problem 4.71   (a) What is the radius of the largest circle that can fit in a $120°$ sector of a circle of radius 1?

   Solution: $2\sqrt{3} - 3$.

   Consider the general diagram from 4.66. We again have a 30-60-90 triangle, but now the hypotenuse has length $2r/\sqrt{3}$. Therefore, $1 = \dfrac{2r}{\sqrt{3}} + r$ so $r = \dfrac{1}{1 + 2/\sqrt{3}} = 2\sqrt{3} - 3$.

  (b) What if you have a $240°$ sector?

   Solution: $1/2$.

   Note that the general diagram does *not* apply, as we are limited by the center of the circle. The largest possible radius is one half of the original radius, so $1/2$.

Problem 4.72 Let $\angle APB$ be an inscribed angle on a circle with center O. Prove that $\angle APB$ is half the angular size of arc $\widehat{AB}$ if:

  (a) O lies on $\angle APB$.

   Solution: Assume O lies on $\overline{AP}$. Then $\triangle OPB$ is isosceles, so $2\angle BPO = 2\angle BPA = \angle BOA$ as needed.

  (b) O lies inside $\angle APB$.

   Solution: Let Q be such that $\overline{PQ}$ is a diameter. Note $\triangle BOP$ is isosceles, so $\angle BOQ = 2\angle BPO$. Similarly, $\angle AOQ = 2\angle APO$. Hence, $\angle BPA = \angle BPO + \angle APO = (\angle BOQ + \angle AOQ)/2 = \angle BOA/2$ as needed.

Problem 4.73 Prove that if two chords AC, BD intersect inside a circle at point P then the measure of $\angle APB$ is half the sum of the angular sizes of $\overarc{AB}, \overarc{CD}$.

Solution: Draw $\overline{BC}$. We have $\angle APB = \angle DBC + \angle ACB$ equals have the sum of the angular sizes of $\overarc{CD}, \overarc{AB}$ (as $\angle DBC, \angle ACB$ are inscribed angles).

Problem 4.74 Suppose $\overarc{AB}$ is an arc with angular size $60°$ and CD is a diameter such that if rays $\overrightarrow{BA}, \overrightarrow{DC}$ are extended to intersect at a point E, $\angle AEC = 30$. Find the angular size of arc $\overarc{BD}$.

Solution: $90°$.

Let the angular measure of $\overarc{BD} = x$. Then the size of $\overarc{AC} = 180 - 60 - x = 120 - x$. Then $\angle AEC = 30 = \dfrac{x - (120 - x)}{2}$ so solving for x gives $x = 90°$.

Problem 4.75 Suppose two perpendicular chords intersect and divide each other in a ratio of $1 : 2$. Find the radius of the circle if each chord is 12in long.

Solution: $2\sqrt{10}$.

The two chords divide each other in segments of 4 and 8 inches. The perpendicular bisector of each chord goes through the center of the circle, so we can form a right triangle with sides 2 and 6 with hypotenuse r, the radius of the circle, as in the diagram below.

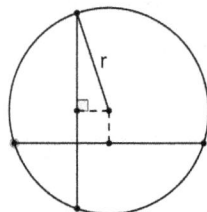

Hence, $r = \sqrt{2^2 + 6^2} = 2\sqrt{10}$.

Problem 4.76 Suppose ω is a circle with radius 6 and center O. Let $\overarc{AB} = 135°$. Let C be on ω such that $\overline{OA} \parallel \overline{BC}$. Find $[OACB]$.

Solution: $18 + 9\sqrt{2}$.

Let E on $\overline{BC}$ be such that $\overline{OE}$ is a height, as in the following diagram.

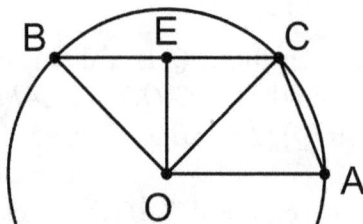

As $\overline{OE} \perp \overline{BC}$ and O is the center of the circle, E is the midpoint of $\overline{BC}$. Combined with the fact that $\overline{OA} \| \overline{BC}$, both $\triangle OEB, \triangle OEC$ are 45-45-90 triangles. Hence, $OA = OB = 6$ so $BE = EO = 6/\sqrt{2} = 3\sqrt{2}$ and $BC = 6\sqrt{2}$. Thus, the trapezoid has bases $6, 6\sqrt{2}$ and height $3\sqrt{2}$, so it has area $\frac{1}{2} \cdot (6 + 6\sqrt{2}) \cdot 3\sqrt{2} = 18 + 9\sqrt{2}$.

Problem 4.77 Suppose you have a regular hexagon with side length 6. Let G, H, I be midpoints of every other side of the hexagon. Let $\overarc{GH}, \overarc{HI}, \overarc{IG}$ be arcs inside the hexagon of three non-overlapping circles with radii 9. Find the area of the region in the hexagon in between all three arcs.
Solution: $81\sqrt{3} - 81\pi/2$.
Note the centers of the three circles form an equilateral triangle with side length $9 + 9 = 18$, as in the diagram below.

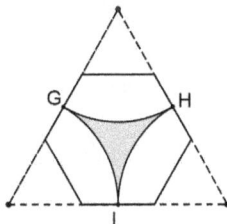

Hence, the area we want is the area of this triangle minus three $60°$ sectors (of circles with radius 9). Hence the area is $18^2 \cdot \frac{\sqrt{3}}{4} - 3 \cdot \frac{1}{6} \cdot \pi 9^2$.

Problem 4.78 Let $\angle APB$ be an inscribed angle on a circle with center O. Prove that $\angle APB$ is half the angular measure of arc $\overarc{AB}$ if O lies outside $\angle APB$.
Solution: Let Q be such that $\overline{PQ}$ is a diameter. Note $\triangle BOP$ is isosceles, so $\angle BOQ = 2 \cdot \angle BPO$. Similarly, $\angle AOQ = 2 \cdot \angle APO$. Hence, $\angle BPA = \angle APO - \angle BPO = (\angle AOQ - \angle BOQ)/2 = \angle AOB/2$ as needed.

Problem 4.79 Prove that if two chords AC, BD intersect outside a circle at point P then the measure of $\angle APB$ is half the difference of the angular sizes of $\overarc{AB}, \overarc{CD}$.
Solution: Let the points be as in the diagram earlier (so C, D are on $\overline{AP}, \overline{BP}$). Connect $\overline{BC}$. Then using $\triangle ACB$ we have $\angle APB + \angle CBD = \angle ACB$. Therefore, $\angle APB = \angle ACB - \angle CBD$. Since $\angle ACB, \angle CBD$ are inscribed angles with respective arcs $\overarc{AB}, \overarc{CD}$, the result follows.

Problem 4.80 Suppose A, B, C are points on a circle such that the angular measures of arc $\overarc{AB}$ (not containing C) and arc $\overarc{CA}$ (not containing B) are in ratio $5 : 9$. Suppose further that $\angle ABC = 90°$. Find the measure of $\angle BAC$.

Solution: $40°$.
Since $\angle ABC = 90°$, we have $\widehat{CA} = 180°$. Hence, $\widehat{BC} : \widehat{AB} = 4 : 5$. Hence, $\widehat{BC} = 80°$ and $\angle BAC = 80/2 = 40°$.

Problem 4.81 Suppose $\overline{AB}, \overline{CD}$ are two chords of equal length who intersect at E. Suppose $\angle AED = 120°$, and $AE : EB = CE : ED = 1 : 2$. Further, suppose $AC = 2$.
 (a) Find the distance from E to the center of the circle.

Solution: $\dfrac{2}{\sqrt{3}} = \dfrac{2\sqrt{3}}{3}$.
We have $\angle AEC = \angle BED = 60°$. Further, as the chords have equal length, both $\triangle ACE, \triangle BDE$ are equilateral, with respective side lengths $2, 4$, so $AB = CD = 6$. Let $\overline{AB}, \overline{CD}$ have midpoints F, G respectively, as in the diagram below.

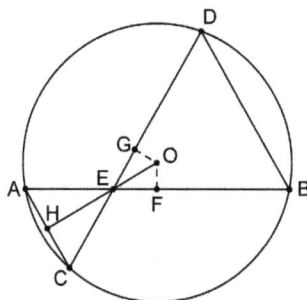

The center of the circle O is such that $\angle OFE = \angle OGE = 90°$ (as the perpendicular bisector of a chord always contains the center). We have $GE = EF = BE - BF = 4 - 3 = 1$. Since $\triangle OGE \cong \triangle OFE$ and thus $\angle OEF = \angle OEG = 60/2 = 60°$, $\triangle OFE$ is a $30 - 60 - 90$ triangle. This implies that EO has length $2/\sqrt{3} = 2\sqrt{3}/3$ which is the distance between E and the center of the circle.

 (b) Find the radius of the circle.
Solution: $\sqrt{28/3}$.
We continue from part (a). Let H be the midpoint of $\overline{AC}$. Then $\triangle OHA$ is a right triangle with hypotenuse OA the radius of the circle. We have (since H is a midpoint) $AH = 1$, and $OH = OE + EH = 2/\sqrt{3} + \sqrt{3} = 5/\sqrt{3}$ (since $\overline{EH}$ is the height of an equilateral triangle with side length 2). Using the Pythagorean Theorem we then find $OA^2 = 28/3$, so the radius is $\sqrt{28/3}$.

Problem 4.82 Recall the construction of trapezoid $OACB$ as in Problem 4.76 What angle does $\widehat{AB}$ have to be so that $ABCD$ is a parallelogram?
Solution: $120°$.
For $OACB$ to be a parallelogram, it must break into two equilateral triangles, hence $\angle AOB = 120°$.

Problem 4.83 (a) Suppose two inscribed angles share the same arc. Prove that the angles are equal.

Solution: Both angles are equal to half the angular size of the arc.

(b) Prove the Power of a Point formula for chords intersecting inside the circle.
Solution: We have intersecting chords as below.

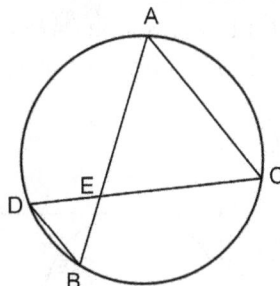

Note if we show $\triangle AEC \sim \triangle DEB$ then $\dfrac{AE}{CE} = \dfrac{DE}{BE}$ so $AE \cdot BE = CE \cdot DE$ as needed. We have $\angle AEC, \angle DEB$ are vertical. We also have $\angle CAE = \angle CAB = \angle CDB = \angle EDB$ as they share the same arc (similarly, $\angle ACE = \angle DBE$) so the triangles are similar and the result follows.

Problem 4.84 Suppose a line is tangent to the circle at C and it intersects a diameter $\overline{AB}$ at E. Prove that $CE^2 = AE \cdot BE$.

Solution: Suppose O is the center of the circle, and B lies between A, E, as in the diagram below.

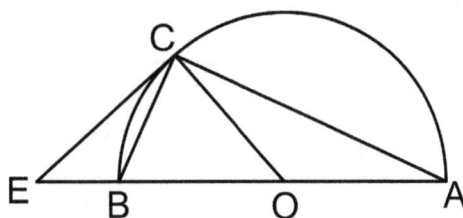

As in the previous problem, it suffices to show $\triangle AEC \sim \triangle CEB$, as then after cross multiplying, $AE \cdot BE = CE \cdot CE$. Both triangles share the angle $\angle CEB = \angle CEA$. Since $\overline{CE}$ is tangent at C, $\angle OCE = 90°$. Since $\overline{AB}$ is a diameter (so $\overarc{AB}$ has angular size $180°$), $\angle ACB = 90°$. Hence $\angle ACO = 90 - \angle BCO = \angle BCE$. Further, $\angle ACO = \angle CAO$ (as $\triangle ACO$ is isosceles) and $\angle CAE = \angle CAO = \angle BCE$ so $\triangle AEC \sim \triangle CEB$ as needed.

Problem 4.85 Suppose chords $\overline{AB}, \overline{CD}$ intersect at E, such that $AE : EB = 1 : 3$ and $CE : ED = 1 : 12$. Find the ratio of $AB : CD$.

Solution: $\dfrac{8}{13}$.

Let $AE = x, CD = y$, so $EB = 3x, ED = 12y$. By Power of a Point, $3x^2 = 12y^2$, so solving for x/y we get $x/y = 2$. Hence, $\dfrac{AB}{CD} = \dfrac{4x}{13y} = \dfrac{4}{13} \cdot 2 = \dfrac{8}{13}$.

Problem 4.86 Suppose we have a rectangle $ABCD$ with $AB = 8$, $BC = 12$. "Inscribe" a circle in the rectangle so that it touches sides $\overline{AB}, \overline{BC}, \overline{AD}$. Let M be the midpoint of $\overline{AB}$. Call $E \neq M$ the intersection of $\overline{MD}$ with the circle. Find DE.

Solution: $8\sqrt{10}/5$.

Consider the diagram below.

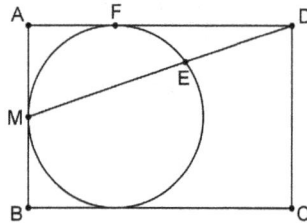

Let F be such that $\overline{AD}$ is tangent to the circle at F. Since $AB = 8$, the radius of the circle is 4. Hence, $AM = 4$ and $DF = 12 - 4 = 8$. By Power of a Point, $DF^2 = DM \cdot DE$. As $DM = \sqrt{4^2 + 12^2} = 4\sqrt{10}$, $DE = \dfrac{8^2}{4\sqrt{10}} = \dfrac{8\sqrt{10}}{5}$.

Problem 4.87 Prove the Power of a Point formula for two chords whose extensions intersect outside a circle.

Solution: Consider the diagram below.

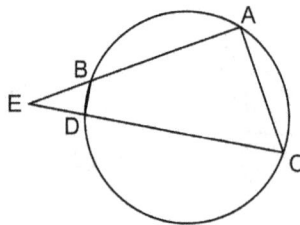

Note that the arcs corresponding to angles $\angle ACD, \angle DBA$ make up the entire circle, so $\angle ACD + \angle DBA = 360/2 = 180°$. Hence, $\angle ACD = 180° - \angle ABD = \angle DBE$. A similar argument gives $\angle CAE = \angle CAB = \angle BDE$. Thus, $\triangle ACE \sim \triangle DBE$, and hence (after cross multiplying) $AE \cdot BE = CE \cdot DE$.

Problem 4.88 Prove the Power of a Point formula for arbitrary chord $\overline{AB}$ which intersects a tangent line (tangent to the circle at C) at E.

Solution: Let $\overline{DF}$ be a diameter whose extension contains E. By Problem 4.84 $CE^2 = DE \cdot FE$. By the previous question, $DE \cdot FE = AB \cdot BE$. Combining these gives the result.

Problem 4.89 Suppose diameter $\overline{CD}$ intersects chord $\overline{AB}$ at E, so that $AE = 4, EB = 9$. If the diameter of the circle is 15, Find CE and ED.
Solution: $3, 12$.
Using Power of a Point we have $AE \cdot EB = CE \cdot ED$, so $CE \cdot ED = 36$. We also have $CE + ED = 15$, from here we can solve for CE, ED.

Problem 4.90 Recall the setup from Problem 4.86 Assume that CE is tangent to the circle at E (It is, as a challenge try to prove it!). Find the length of CE.
Solution: 8.
Suppose $\overline{CM}$ intersects the circle at $E' \neq M$. By symmetry, $DE = CE', DM = CM$. By Power of a Point, $CE^2 = CE' \cdot CM = DE \cdot DM$. As in 4.86 $DE \cdot DM = 64$, so $CE = 8$.

Problem 4.91 Inscribing and Circumscribing a Sphere
  (a) Find the volume of the largest sphere that fits in a cube of volume 1. (That is, inscribe a sphere inside the cube.)
  Solution: $\dfrac{1}{6}\pi$.
  The diameter of the sphere must be 1, so the radius is $1/2$. Hence the volume is
  $$\frac{4\pi}{3}\left(\frac{1}{2}\right)^3 = \frac{\pi}{6}.$$

  (b) Find the volume of the smallest sphere that holds a cube of volume 1. (That is, circumscribe a sphere outside the cube.)
  Solution: $\dfrac{\sqrt{3}}{2}\pi$.
  The diameter of the sphere must be $\sqrt{3}$ so the radius is $\sqrt{3}/2$. Hence the volume is $\dfrac{4\pi}{3}\left(\dfrac{\sqrt{3}}{2}\right)^3 = \dfrac{\pi\sqrt{3}}{2}$.

Problem 4.92 (2010 AMC 10A) Suppose we have a cube with side length 4. In the middle of each face of the cube, cut a 2 by 2 square hole all the way through the cube. What is the volume of the remaining solid after all the holes are cut.
Solution: 32.
All the holes cut out right rectangular prisms with dimensions 2 by 2 by 4, so have volume $2 \cdot 2 \cdot 4 = 16$. There are three such holes. However, each hole goes through a central cube with side length 2 of volume $2^3 = 8$. Hence the removed volume is

$3 \cdot 16 - 2 \cdot 8 = 32$. Hence the remaining volume of the solid is $4^3 - 32 = 32$.

Problem 4.93 Suppose you have a regular tetrahedron with side length a.
  (a) Show that the surface area is $a^2\sqrt{3}$.
    Solution: A regular tetrahedron is made up of 4 equilateral triangles. Each of these has area $a^2\sqrt{3}/4$ so the total surface area is $a^2\sqrt{3}$.

  (b) Show that the volume is $\dfrac{a^3\sqrt{2}}{12}$.

    Solution: Recall a tetrahedron is triangular pyramid, so have volume $\frac{1}{3}Bh$. The base has area $a^2\sqrt{3}/4$, so we are left to find the height. Since we have a regular tetrahedron, the apex is above the incenter of the base. Recall for an equilateral triangle with side length a, the incenter is $a/\sqrt{3}$ from the vertex. This leads to the following diagram:

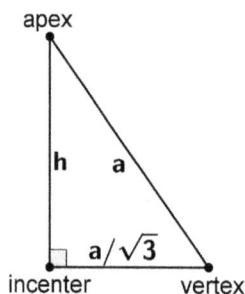

By the Pythagorean Theorem, $a^2 = h^2 + (a/\sqrt{3})^2$, so solving for h we have $h = a\sqrt{2/3}$. Thus, the volume is $\dfrac{1}{3} \cdot \dfrac{a^2\sqrt{3}}{4} \cdot \dfrac{a\sqrt{2}}{\sqrt{3}} = \dfrac{a^3\sqrt{2}}{12}$ as needed.

Problem 4.94 Suppose $S-ABC$ is a regular tetrahedron with apex S. Cut off the top "half" of the tetrahedron (that is, cut through the midpoints of $\overline{SA}, \overline{SB}, \overline{SC}$ and leave the bottom).
  (a) How many vertices, edges, and faces does the resulting solid have?
    Solution: $6, 9, 5$.
    1 vertex is removed, but 3 new are added, so there are $4 - 1 + 3 = 6$ vertices. 3 new edges are added, so there are $6 + 3 = 9$ edges. 1 new face is added, resulting in $4 + 1 = 5$ faces.

  (b) Find the volume and the surface area as ratios to the original volume and surface area. Hint: Try to do so without using Problem 4.93.
    Solution: $7 : 8$ for both.
    Note the removed part of the original tetrahedron is itself a tetrahedron, with half

the side length of the original. From here it follows (using similar tetrahedrons) that the volume removed is $1/8$ of the original volume. Similar reasoning gives that each face of the removed tetrahedron has area $1/16$ of the surface area of the original tetrahedron (using similar triangles). As we are removing 3 and adding 1 of these faces, the net change in surface area is also $3/16 - 1/16 = 1/8$th.

Problem 4.95 (2014 AMC 10A) Stack cubes with side length $1, 2, 3, 4$ as shown in the diagram below.

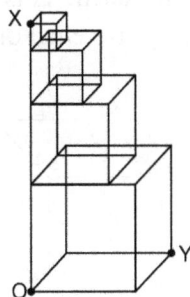

(a) Find the distance from X to Y.
   Solution: $2\sqrt{33}$.
   Using the distance formula we have the distance is $\sqrt{10^2 + 4^2 + 4^2} = 2\sqrt{33}$.

(b) Find the length of the portion of $\overline{XY}$ contained in the bottom square.
   Solution: $4\sqrt{33}/5$.
   It may help to view the cross section formed by the plane containing points X, O, Y:

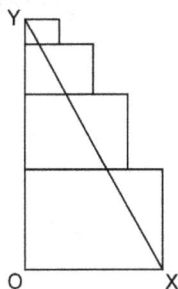

   First note that the entire line $\overline{XY}$ is contained inside the cubes. Form a triangle similar to $\triangle XOY$ with hypotenuse of length x equal to the portion of $\overline{XY}$ in the bottom square. Since this triangle has height 4, we have $\dfrac{2\sqrt{33}}{10} = \dfrac{x}{4}$, so solving for x we get $x = \dfrac{4\sqrt{33}}{5}$.

Problem 4.96 (2010 AMC 10A) Suppose a bored bee lives on a cube with side length 1. For "fun" he decides to visit every vertex of the cube, each exactly once, starting and ending at the same vertex. It will travel from one vertex to another using straight lines (either crawling or flying). Give an example of a path that uses the maximum distance and find this distance.

Solution: $4\sqrt{2}+4\sqrt{3}$.

The distance between opposite corners of the cube is $\sqrt{3}$ and the distance opposite corners of a square face is $\sqrt{2}$. Note any path has 8 stops, and it is possible to travel only using the above distances, 4 times each. This is optimal, as there are only 4 pairs of opposite corners (and the bee can only use each once). An example path is given below:

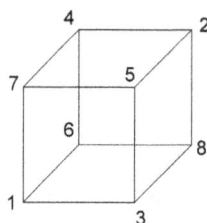

(The bee travels from 1 to 2 to 3 ... to 8 to 1.)

Problem 4.97   (a) Four identical balls (spheres), each of radius 1in, are glued to the ground so that their centers form the vertices of a square with side length 2in. Suppose you rest a fifth identical ball on the four balls (so the fifth ball is a sphere externally tangent to the other spheres). How far does this ball rest off the ground?

Solution: $\sqrt{2}$.

Suppose the balls have centers A, B, C, D and E (for the fifth sphere on top). The view from the top is as below:

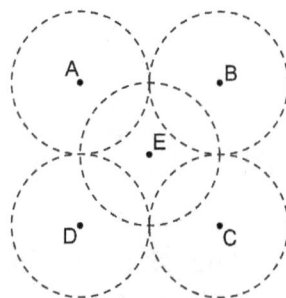

Since $ABCD$ is a square, we have that the diagonal AC is $2\sqrt{2}$, and thus, by symmetry, $AE = CE = \sqrt{2}$. Looking at the cross section in the plane formed by $\triangle ACE$ we have:

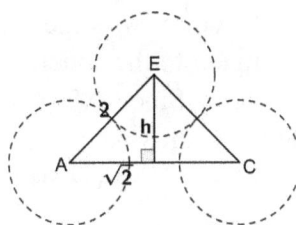

Let $h+1$ denote the height of the center of the fifth sphere off the ground. Note that since all spheres have radius 1in, this means the sphere rests $h+1-1=h$ inches off the ground. Setting up right triangles gives us $2^2 = 2+h^2$, so solving for h gives $\sqrt{2}$.

(b) Repeat the setup of (a). Suppose now, however, that the fifth ball is not the same size as the others, and that it rests 1in off the ground. Find the radius of the fifth ball.

Solution: $1/2$.

Proceed similarly to part (a), yielding the following picture.

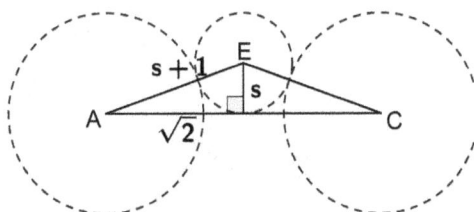

Let s denote the radius of the last sphere. Setting up right triangles again gives us $(s+1)^2 = s^2+2$. Solving for s gives $s = 1/2$.

Problem 4.98 Spheres in a Cone

(a) Find the volume of the largest sphere that can fit inside a cone of radius 1 and height $\sqrt{3}$.

Solution: $\dfrac{4\sqrt{3}}{27}\pi$.

By symmetry, the cross section will look like the diagram below:

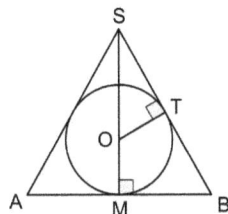

where S is the apex of the cone, M is the center of the base, O the center of the sphere, and T a point tangent to the sphere. Since $AM = BM = 1$ and $SM = \sqrt{3}$, $\triangle SMB$ is a 30-60-90 triangle. Since $\triangle STO$ is a right triangle, it is also a 30-60-90 triangle. Let r denote the radius of the sphere. We have $TO = r$, so $SO = 2r$. Hence $2r + r = SO + OM = SM = \sqrt{3}$ so $r = \sqrt{3}/3$. Hence the volume of the sphere is $\dfrac{4\pi}{3}\left(\dfrac{\sqrt{3}}{3}\right)^3 = \dfrac{4\pi\sqrt{3}}{27}$.

(b) Assume the sphere as in (a) is placed inside the cone. Suppose you now want to fit another sphere in the cone that is tangent to the base. Find the radius of the largest such sphere.

Solution: $\dfrac{\sqrt{3}}{9}$.

Consider a "zoomed-in" diagram that extends the diagram in (a) as below.

Here P is the center of the second sphere with radius s. Note that $\triangle BMO \sim \triangle PNO$ (AA), so since $OM : BM = 1 : \sqrt{3}$, we have $\triangle PNO$ is a 30-60-90 triangle, so $NO : PO = 1 : 2$. However, we know $NO = r - s, PO = s + r$, where $r = \sqrt{3}/3$ is the radius of the sphere calculated in part (a). Solving $2(r - s) = s + r$ we get $s = r/3$ so $s = \sqrt{3}/9$.

Problem 4.99 Find the surface area of the remaining solid in Problem 4.92

Solution: 144.

We use a method similar to the method used to calculate the remaining volume. The original cube has surface area $6 \cdot 4^2 = 96$. Each hole has a "surface area" of $2 \cdot 2^2 + 4 \cdot (2 \cdot 4) = 144$. However, we have counted the missing 6 squares on the faces of the original cube twice each, so we over counted the surface area by $2 \cdot 6 \cdot 2^2 = 48$. We have also counted the surface area of the missing cube at the center of the original cube twice (double check this!), so we also over counted the surface area by $2 \cdot 6 \cdot 2^2 = 48$. Hence, the total surface area is $96 + 144 - 48 - 48 = 144$.

Problem 4.100 Suppose you pick 4 vertices of a cube to form a tetrahedron.

(a) How many different (non-congruent) tetrahedrons are possible? Are any of them regular tetrahedrons?

Solution: 4.

Label the cube $ABCD - A'B'C'D'$. First assume, 3 of the vertices are contained in the same square face. Hence, without loss of generality, the tetrahedron consists of vertices A, B, C, and one of A', B', C', D'. Note $ABC - B' \cong ABC - C'$, so this leads to 3 different tetrahedron, none of which are regular.

The only other possibility (up to congruence) is (again using the labeling $ABCD - A'B'C'D'$) $AC - B'D'$. Note this tetrahedron *is* regular.

(b) Find the volumes for each of the possibilities in (a) if the cube has volume 1.
Solution: $1/6$ or $1/3$.

If 3 vertices are contained in the same square face, then we have a tetrahedron (which is a triangular pyramid) with base of area $1/2$ and height of 1. Thus, the volume is $\frac{1}{3} \cdot \frac{1}{2} \cdot 1 = \frac{1}{6}$.

Now suppose we have tetrahedron $AC - B'D'$. Note this tetrahedron divides the rest of the cube into 4 congruent tetrahedrons, which we just saw have volume $\frac{1}{6}$.

Hence, tetrahedron $AC - B'D'$ has volume $\frac{1}{3}$.

Problem 4.101 Suppose you have a regular square pyramid $S - ABCD$ with height 6 whose square has side length 4. Call the midpoints of the square E, F, G, H (on $\overline{AB}, \overline{BC}, \overline{CD}, \overline{DA}$ respectively) and the midpoints of $\overline{SA}, \overline{SB}, \overline{SC}, \overline{SD}$ respectively T, U, V, W. Form polyhedron $EFGH - TUVW$ (with 10 faces).

(a) Describe the faces of $EFGH - TUVW$. How many vertices and edges does the polyhedron have?
Solution: Consider the diagram below.

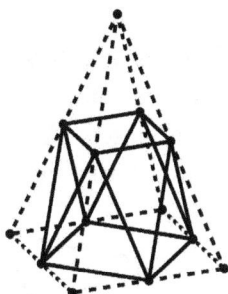

There are two square faces (both different sizes, with $TUVW$ rotates $45°$ compared to $EFGH$), and 8 triangular faces (2 groups of 4 congruent triangles). There are 8 vertices and 16 edges.

(b) Find the volume of $EFGH - TUVW$. Hint: Do this indirectly.

Solution: 20.

Note the pyramid $S-ABCD$ is made up of (i) the polyhedron $EFGH-TUVW$, (ii) the regular square pyramid $S-TUVW$, and (iii) four congruent triangular pyramids ($T-AEH, U-BFE, V-CGE, W-DGH$). The volume of the entire pyramid is $\frac{1}{3}\cdot 4^2 \cdot 6 = 32$. The volume of the smaller square pyramid (ii) is $\frac{1}{3}\cdot 2^2 \cdot 3 = 4$ (using similar triangles). Lastly, the volume of each triangular pyramid (iii) is $\frac{1}{3}\cdot\frac{1}{2}\cdot 2^2 \cdot 3 = 2$. Hence, the volume (i) we want is $32-4-4\cdot 2 = 20$.

Problem 4.102 Suppose solid cubes of side length 1 are removed from every corner of a solid cube with side length 3.

(a) What is the volume and surface area of the new solid?
Solution: $19, 54$.
The volume is $3^3 - 8\cdot 1^3 = 19$ and the surface area is unchanged (think of "folding in" the corners), so it is $6\cdot 3^2 = 54$.

(b) How many vertices, edges, and faces does the new solid have?
Solution: $56, 84, 30$.
In every corner we: remove the original vertex and add 7 new vertices, add 9 new edges, and add 3 new faces. Hence, the new solid has $8\cdot 7 = 56$ vertices, $12+8\cdot 9 = 84$ edges, and $6+8\cdot 3 = 30$ faces.

Problem 4.103 Suppose you have a unit cube. Pick two opposite corners. In each corner, form a tetrahedron using the corner and the three adjacent vertices. Remove these two tetrahedrons and call the resulting polyhedron $\mathscr{S}$.

(a) How many vertices, edges, and faces does the resulting polyhedron have? Describe the faces.
Solution: 6 vertices, 12 edges, and 8 faces.
Note we are removing 2 vertices, not changing the number of edges, and adding 2 faces. All of the faces are triangles. The following diagram gives a rough idea of what the resulting polyhedron looks like.

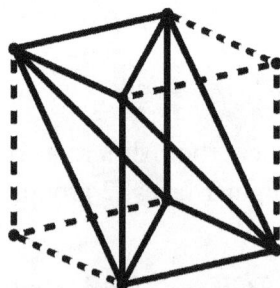

(b) Find the volume of $\mathscr{S}$.
   Solution: 2/3.
   Note each of the tetrahedrons removed have base area 1/2 and height 1, hence has volume 1/6. Hence $\mathscr{S}$ has volume $1 - 2 \cdot \dfrac{1}{6} = \dfrac{2}{3}$.

(c) Suppose $\mathscr{S}$ is resting on one of the faces (ignore whether the polyhedron would actually balance or not). What are different possible "heights" of $\mathscr{S}$?
   Solution: 1 or $\sqrt{3}/3$.
   First note that if it is resting on one of the faces that was part of an original face of the cube, the height is still 1. The original cube diagonal connecting the opposite corners has length $\sqrt{3}$, if we let h denote the height of each removed tetrahedron (measured along this diagonal), then the height of $\mathscr{S}$ is $\sqrt{3} - 2h$. Along this height, the tetrahedron's base is an equilateral triangle with side length $\sqrt{2}$, hence has area $\sqrt{3}/2$. Then (as we calculated in part (b)) $\dfrac{1}{6} = \dfrac{1}{3} \cdot \dfrac{\sqrt{3}}{2} h$ so $h = \dfrac{\sqrt{3}}{3}$. Thus, $\mathscr{S}$ has height $\sqrt{3} - 2 \cdot \dfrac{\sqrt{3}}{3} = \dfrac{\sqrt{3}}{3}$.

Problem 4.104 Stack cubes with side length $1,2,3,4$ as shown in the diagram below.

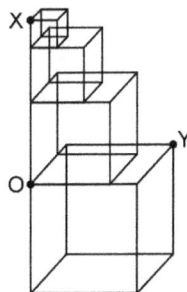

(a) Find the distance from X to Y.
   Solution: $2\sqrt{17}$.
   Using the distance formula we have the distance is $\sqrt{6^2 + 4^2 + 4^2} = 2\sqrt{17}$.

(b) Find the length of the portion of $\overline{XY}$ *not* contained inside the cubes.
   Solution: $\dfrac{\sqrt{17}}{2}$.
   Consider the diagram in the plane containing X, O, Y:

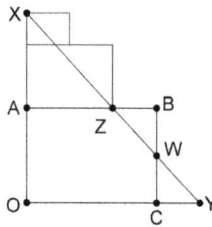

First note that $\triangle XOY \sim \triangle XAB$ with ratio of sides $2:1$, so $AZ = OY/2 = 4\sqrt{2}/2 = 2\sqrt{2}$ (note this shows that Z is actually on the corner of the cube as in the diagram) as well as $XZ = XY/2$. We then know $\overline{AB} \parallel \overline{OY}$, which implies $\triangle ZBW \sim \triangle YCW$. Since $BZ = AB - AZ = 3\sqrt{2} - 2\sqrt{2} = \sqrt{2}$ and $CY = OY - OC = 4\sqrt{2} - 3\sqrt{2} = \sqrt{2}$, the ratio of sides is $1:1$, so $WY = ZW = YZ/2 = XY/4$. Since $\overline{WY}$ is the portion of $\overline{XY}$ outside the cubes, our answer is $2\sqrt{17}/4 = \sqrt{17}/2$.

Problem 4.105   (a) Repeat Problem 4.96 if the bee wants the shortest path.
Solution: 8.
There are many examples of paths that only travel on the 8 edges of the cube, which is clearly the shortest path.

(b) Find the maximum and minimum paths if the bee instead lives on a Right Prism with height 2 whose base is a regular hexagon with side length 1.
Solution: Max: $6\sqrt{7} + 6\sqrt{5}$, Min: 14.
Inside the hexagonal bases, it is possible for the bee to travel distances of 1, 2, or $\sqrt{3}$ as in the diagram below:

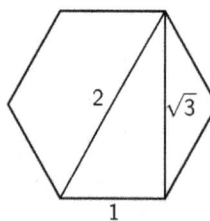

Combined with a height of 2, this gives us paths from one base to the other base of (using the Pythagorean Theorem): $\sqrt{2^2 + 1^2} = \sqrt{5}, \sqrt{2^2 + 2^2} = 2\sqrt{2}, \sqrt{2^2 + \sqrt{3}^2} = \sqrt{7}$. Since there are 12 vertices, any path will have 12 steps. For the maximum path, note it is always best to alternate between the bases. Since we cannot backtrack, it is impossible to have two $\sqrt{7}$ length steps in a row. It is, however, possible to have a path that alternates between paths of length $\sqrt{7}$ and $\sqrt{5}$ (so 6 of each), giving a maximum path of $6\sqrt{7} + 6\sqrt{5}$. One such path is given below:

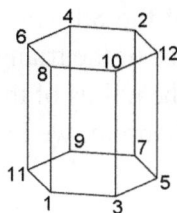

For the minimum, we want to alternate bases as little as possible (two times is the minimum). Hence a minimum path has 10 steps of length 1, and two steps of length 2. An example of one of these paths is given below:

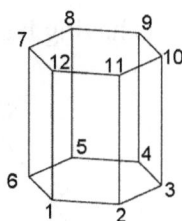

which has the minimum length of $10 + 2 \cdot 2 = 14$.

Problem 4.106 (2013 AMC 10A) Suppose 6 spheres of radius 1 are arranged so that the centers form a regular hexagon with side length 2. All 6 spheres are internally tangent to a larger 7th sphere whose center is the center of the hexagon. Lastly, an 8th sphere is externally tangent to the 6 smaller spheres and internally tangent to the large sphere.

(a) Find the radius of the large sphere.

Solution: 3.

Recalling that a regular hexagon can be divided into 6 equilateral triangles, the distance between opposite vertices in the hexagon is 4. Hence the diameter of the 7th sphere is $1 + 4 + 1 = 6$.

(b) Find the radius of the 8th sphere.

Solution: $3/2$.

We proceed similarly to Problem 4.97. The side view we get is as in the diagram below.

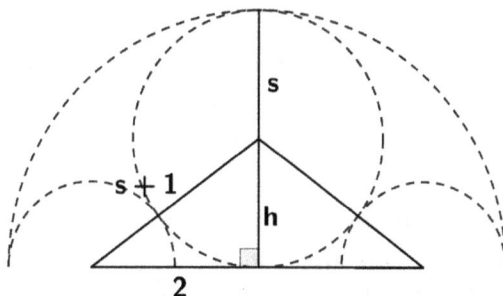

Let s denote the radius of the 8th sphere. Note we cannot assume that the 8th sphere is tangent to the base of the triangle in the diagram, so we can only say that $h = 3 - s$, where the 3 is the radius of the large 7th sphere we calculated in part(a). Using the Pythagorean Theorem, we have $(s+1)^2 = 2^2 + (3-s)^2$. Solving for s gives $s = 3/2$.

Problem 4.107 Suppose you have tetrahedron $S - ABC$. Cut the tetrahedron with a plane between S and $\triangle ABC$, forming a new tetrahedron $S - A'B'C'$ (with A' on $\overline{SA}$, etc.). Prove that the ratio of volumes of $S - ABC$ to $S - A'B'C'$ is equal to $SA \cdot SB \cdot SC :$ $SA' \cdot SB' \cdot SC'$.

Solution: View the tetrahedrons as having bases $\triangle SAB$ and $\triangle SA'B'$, as in the diagram below.

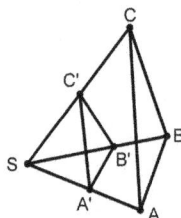

Let h, h' be the corresponding heights. Using similar triangles we have $h : h' = SC : SC'$. Further, since triangles $\triangle SAB, \triangle SA'B'$ share an angle, we have $[SAB] : [SA'B'] = SA \cdot SB :$ $SA' \cdot SB'$. As the volume of a tetrahedron is one third the base area times the height, the result follows.

Problem 4.108 (2015 AMC 10B #17) Suppose you have a right rectangular prism with length, width, height equal to $3, 4, 5$. Connect all the centers of the faces to form an octahedron (a polyhedron with 8 sides). What is the octahedron's volume?
Solution: 10.
Use the diagram below to helpful visualize the problem.

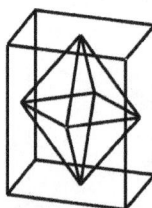

Cut the right rectangular prism in half to form two congruent right rectangular prisms with dimensions $3, 4, 5/2$. This also cuts the octahedron into two congruent pyramids. Each pyramid has a height of $5/2$ and a base that is a rhombus with diagonals $3, 4$.

Recalling that a rhombus has area equal to half the product of its diagonals, the base of each of our pyramids has area $\dfrac{3\cdot 4}{2} = 6$. The total volume of the octahedron is thus
$$2\times\left(\frac{1}{3}\cdot 6\cdot\frac{5}{2}\right) = 10.$$

Problem 4.109 Suppose you start with a right cone and cut off the top of the cone with a plane parallel to the base. The resulting solid, called a *frustum* has two circular "bases", say with radii R and r (with $R > r$), and height H. (Hence from the side the frustum looks like a trapezoid with bases R, r and height H.)

(a) Show that the volume of a frustum is $\dfrac{\pi H}{3}(R^2 + Rr + r^2)$.

Solution: Let $H + h$ denote the height of the original cone (see the side cross section below).

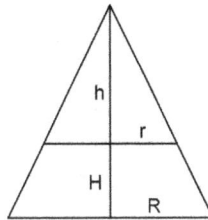

The volume of the frustum thus the difference between a cone of height $H + h$ and radius R and a cone of height h and radius r. Thus the volume of the frustum is $\pi R^2(H+h)/3 - \pi r^2 h/3$. Using similar triangles we have $\dfrac{h}{r} = \dfrac{h+H}{R}$ so solving for h we get $h = \dfrac{rH}{R-r}$. Substituting into our volume equation and simplifying gives the above result.

(b) Suppose a sphere can be inscribed in a frustum with base radii r, R such that the sphere is tangent to the two bases and the side. Find the radius of such a sphere in terms of r, R.
Solution: $\sqrt{rR}$.
Let s denote the radius of the sphere. We have the following diagram (another side view):

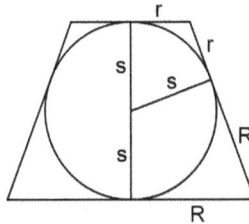

Therefore, we must have $H = 2s$ for the sphere to be tangent. Further, dropping a height from an endpoint of the top base yields a right triangle, so $(R+r)^2 = (2s)^2 + (R-r)^2$. Solving for s gives $s = \sqrt{rR}$.

Problem 4.110 Start with right triangular prism $ABC - DEF$. Divide the prism into four parts using the planes through points A, B, F and D, E, C. Find the ratio of the volumes of these four parts.

Solution: $1 : 3 : 3 : 5$.

Suppose the entire volume is 1. Let P and Q be respectively the intersections of $\overline{AF}, \overline{CD}$ and $\overline{BF}, \overline{CE}$. Since we have a right prism, $ACFD$ and $BCFE$ are rectangles, so P and Q are midpoints of all the respective segments (as the diagonals of a rectangle bisect each other). The following diagram is helpful for reference.

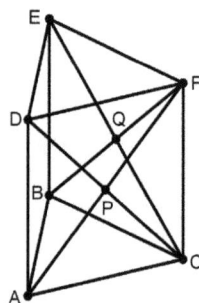

The volume of $F - ABC$ (a triangular prism) is $1/3$ (and similarly so is the volume of $C - DEF$). Using Problem 4.107, the volume of tetrahedron $F - PQC$ is $(1/2)^2 = 1/4$ of this volume, hence $1/12$. Two of the remaining regions thus have volume $1/3 - 1/12 = 1/4$ (the regions formed by removing tetrahedron $F - PQC$ from triangular prisms $F - ABC$ and $C - DEF$). Therefore the last region has area $1 - /12 - 1/4 - 1/4 = 5/12$. Therefore, the ratio of volumes is $1 : 3 : 3 : 5$.

Problem 4.111 Suppose you have a right pyramid with a regular hexagon as a base. Suppose further that the surface area of the pyramid is 3 times the area of the base. If the volume of the pyramid is $6\sqrt{3}$, find side length of the hexagon.

Solution: 2

Let a denote the side length of the hexagon. Therefore, the area of the hexagon is $3a^2\sqrt{3}/2$ and the inradius squared $r^2 = 3a^2/4$. (Recall a hexagon can be divided into 6 equilateral triangles. The inradius is just the altitude of these triangles.) Hence, if h is the height of the pyramid and L is the slant height, we have $L^2 = 3a^2/4 + h^2$. Now, if the surface area SA of the pyramid is twice the area of the base B, we have $3B = SA = B + PL$ (with $P = 6a$ the perimeter of the hexagon). Hence $2B = PL$ so

$a\sqrt{3} = L$, hence $3a^2 = L^2 = 3a^2/4 + h^2$ and hence $h = 3a/2$. Therefore, the volume of the pyramid $6\sqrt{3} = \dfrac{1}{3} \cdot \dfrac{3a^2\sqrt{3}}{2} \cdot h = \dfrac{1}{3} \cdot \dfrac{3a^2\sqrt{3}}{2} \cdot \dfrac{3a}{2}$ and solving for a yields $a = 2$.

Problem 4.112 Compute the following values.

(a) $\cos 225°$

   Solution: $-\sqrt{2}/2$.
   $\cos 225° = \cos(180° + 45°) = -\cos(45°)$.

(b) $\sin \dfrac{11\pi}{6}$

   Solution: $-1/2$.
   $\sin \dfrac{11\pi}{6} = \sin\left(2\pi - \dfrac{\pi}{6}\right) = -\sin\dfrac{\pi}{6}$.

(c) $\tan 405°$

   Solution: 1.
   $\tan 405° = \tan(360° + 45°) = \tan 45°$.

(d) $\cot \dfrac{9\pi}{8}$

   Solution: $\sqrt{2} + 1$.

   $\cot \dfrac{9\pi}{8} = \cot\left(\pi + \dfrac{\pi}{8}\right) = \cot \dfrac{\pi}{8} = \dfrac{1}{\tan \dfrac{\pi}{8}} = \dfrac{1 + \cos \dfrac{\pi}{4}}{\sin \dfrac{\pi}{4}} = \sqrt{2} + 1$.

(e) $\sin 75°$

   Solution: $(\sqrt{2} + \sqrt{6})/4$.
   $\sin 75° = \sin(45° + 30°)$.

Problem 4.113 Find a formula for:

(a) $\sin(x + 90°)$

   Solution: $\cos x$

(b) $\cos(x + 90°)$

   Solution: $-\sin x$

Problem 4.114 Evaluate the following.

(a) $\cos\left(\dfrac{\pi}{3}\right) \cos\left(-\dfrac{\pi}{12}\right) - \sin\left(\dfrac{\pi}{3}\right) \sin\left(-\dfrac{\pi}{12}\right)$

   Solution: $\dfrac{\sqrt{2}}{2}$.

   $\cos\left(\dfrac{\pi}{3}\right) \cos\left(-\dfrac{\pi}{12}\right) - \sin\left(\dfrac{\pi}{3}\right) \sin\left(-\dfrac{\pi}{12}\right) = \cos\left(\dfrac{\pi}{4}\right) = \dfrac{\sqrt{2}}{2}$.

(b) $\sin\left(2\arccos\dfrac{4}{5}\right)$

Solution: 24/25.

$$\sin\left(2\arccos\frac{4}{5}\right) = 2\sin\left(\arccos\frac{4}{5}\right)\cos\left(\arccos\frac{4}{5}\right) = 2\times\frac{3}{5}\times\frac{4}{5} = \frac{24}{25}.$$

(c) $\cos\left(\arccos\left(-\dfrac{\sqrt{2}}{2}\right) - \dfrac{\pi}{2}\right)$

Solution: $\sqrt{2}/2$.

$$
\begin{aligned}
\cos\left(\arccos\left(-\frac{\sqrt{2}}{2}\right) - \frac{\pi}{2}\right) &= \cos\left(\frac{\pi}{2} - \arccos\left(-\frac{\sqrt{2}}{2}\right)\right) \\
&= \sin\left(\arccos\left(-\frac{\sqrt{2}}{2}\right)\right) \\
&= \frac{\sqrt{2}}{2}.
\end{aligned}
$$

(d) $\cos^4\dfrac{\pi}{24} - \sin^4\dfrac{\pi}{24}$

Solution: $(\sqrt{2}+\sqrt{6})/4$.

$$
\begin{aligned}
\cos^4\frac{\pi}{24} - \sin^4\frac{\pi}{24} &= \left(\cos^2\frac{\pi}{24} + \sin^2\frac{\pi}{24}\right)\left(\cos^2\frac{\pi}{24} - \sin^2\frac{\pi}{24}\right) \\
&= \cos^2\frac{\pi}{24} - \sin^2\frac{\pi}{24} \\
&= \cos\frac{\pi}{12} \\
&= \cos\left(\frac{\pi}{3} - \frac{\pi}{4}\right) \\
&= \frac{\sqrt{2}+\sqrt{6}}{4}.
\end{aligned}
$$

(e) $\cot 70° + 4\cos 70°$

Solution: $\sqrt{3}$.

$$\cot 70° + 4\cos 70° = \frac{\cos 70° + 2\sin 140°}{\sin 70°} = \frac{\cos 70° + 2\sin 40°}{\sin 70°}$$

$$= \frac{\cos 70° + 2\sin(70° - 30°)}{\sin 70°}$$

$$= \frac{\cos 70° + \sqrt{3}\sin 70° - \cos 70°}{\sin 70°}$$

$$= \sqrt{3}.$$

(f) $\cos\dfrac{\pi}{15}\cos\dfrac{2\pi}{15}\cos\dfrac{4\pi}{15}\cos\dfrac{8\pi}{15}$

Solution: $-1/16$.

Let $P = \cos\dfrac{\pi}{15}\cos\dfrac{2\pi}{15}\cos\dfrac{4\pi}{15}\cos\dfrac{8\pi}{15}$, then

$$\sin\dfrac{\pi}{15}\cdot P = \sin\dfrac{\pi}{15}\cos\dfrac{\pi}{15}\cos\dfrac{2\pi}{15}\cos\dfrac{4\pi}{15}\cos\dfrac{8\pi}{15}$$

$$= \frac{1}{2}\sin\dfrac{2\pi}{15}\cos\dfrac{2\pi}{15}\cos\dfrac{4\pi}{15}\cos\dfrac{8\pi}{15}$$

$$= \frac{1}{4}\sin\dfrac{4\pi}{15}\cos\dfrac{4\pi}{15}\cos\dfrac{8\pi}{15}$$

$$= \frac{1}{8}\sin\dfrac{8\pi}{15}\cos\dfrac{8\pi}{15}$$

$$= \frac{1}{16}\sin\dfrac{16\pi}{15}$$

$$= -\frac{1}{16}\sin\dfrac{\pi}{15}.$$

Thus $P = -\dfrac{1}{16}$.

(g) $(1 - \cot 23°)(1 - \cot 22°)$

Solution: 2.

Either convert all to sin and cos, or use the following: $1 = \cot 45° = \cot(22° + 23°) = \dfrac{\cot 22° + \cot 23°}{\cot 22° \cot 23° - 1}$, thus $\cot 23° \cot 22° = \cot 22° + \cot 23° + 1$. Expand the expression to get 2.

(h) $\arctan \dfrac{1}{2} + \arctan \dfrac{1}{3}$

Solution: $\pi/4$.

Let $\alpha = \arctan \dfrac{1}{2}, \beta = \arctan \dfrac{1}{3}$, then $\tan \alpha = \dfrac{1}{2}, \tan \beta = \dfrac{1}{3}$, and

$$\tan(\alpha + \beta) = \frac{\tan \alpha + \tan \beta}{1 - \tan \alpha \tan \beta} = \frac{\dfrac{1}{2} + \dfrac{1}{3}}{1 - \dfrac{1}{2} \cdot \dfrac{1}{3}} = 1,$$

also both $\arctan \dfrac{1}{2}$ and $\arctan \dfrac{1}{3}$ are in $\left(0, \dfrac{\pi}{2}\right)$, hence $\alpha + \beta \in (0, \pi)$, therefore $\alpha + \beta = \dfrac{\pi}{4}$.

(i) $\cos^2 x + \cos^2 \left(x + \dfrac{2\pi}{3}\right) + \cos^2 \left(x + \dfrac{4\pi}{3}\right)$

Solution: $3/2$.

Let

$$
\begin{aligned}
A &= \cos^2 x + \cos^2 \left(x + \frac{2\pi}{3}\right) + \cos^2 \left(x + \frac{4\pi}{3}\right), \\
B &= \sin^2 x + \sin^2 \left(x + \frac{2\pi}{3}\right) + \sin^2 \left(x + \frac{4\pi}{3}\right),
\end{aligned}
$$

then $A + B = 3$, and

$$
\begin{aligned}
A - B &= \cos 2x + \cos \left(2x + \frac{4\pi}{3}\right) + \cos \left(2x + \frac{8\pi}{3}\right) \\
&= \cos 2x - \frac{1}{2} \cos 2x - \frac{\sqrt{3}}{2} \sin 2x - \frac{1}{2} \cos 2x + \frac{\sqrt{3}}{2} \sin 2x \\
&= 0.
\end{aligned}
$$

Thus $A = B = \dfrac{3}{2}$.

Problem 4.115 Simplify:

$$\cos(A + B) \cos B + \sin(A + B) \sin B$$

Solution: $\cos(A + B - B) = \cos(A)$.

Problem 4.116 Simplify:

$$\frac{\cos(180° + t) \sin(t + 360°)}{\sin(-t - 180°) \cos(-180° - t)}$$

Solution: $\dfrac{(-\cos t)\sin t}{\sin t(-\cos t)} = 1..$

Problem 4.117 Simplify:

$$\cfrac{1}{1-\cfrac{1}{1-\cfrac{1}{1-\sin^2 x}}}$$

Solution: $\sin^2 x$.

Problem 4.118 Given that $\csc x + \cot x = 5$, find the value of $\csc x - \cot x$.
Solution: $1/5$.
Note that $(\csc x + \cot x)(\csc x - \cot x) = \csc^2 x - \cot^2 x = 1$, so the answer is $1/5$.

Problem 4.119 Given that $\tan y = -\sqrt{5}$, find the value of $\dfrac{\sqrt{5}\sin y + 2\cos y}{\cos y - \sqrt{5}\sin y}$.
Solution: $-\dfrac{1}{2}$.
Divide top and bottom by $\cos y$:

$$\frac{\sqrt{5}\sin y + 2\cos y}{\cos y - \sqrt{5}\sin y} = \frac{\sqrt{5}\tan y + 2}{1 - \sqrt{5}\tan y} = \frac{-5+2}{1+5} = -\frac{1}{2}.$$

Problem 4.120 Given that $\tan y = -\sqrt{5}$, find the value of $(\sin y - \cos y)^2$.
Solution: $1 + \sqrt{5}/3$.

$$\begin{aligned}
(\sin y - \cos y)^2 &= (\tan y - 1)^2 \cos^2 y \\
&= \frac{(\tan y - 1)^2}{\sec^2 y} \\
&= \frac{(\tan y - 1)^2}{\tan^2 y + 1} \\
&= \frac{(-\sqrt{5} - 1)^2}{5 + 1} \\
&= \frac{6 + 2\sqrt{5}}{6} \\
&= 1 + \frac{\sqrt{5}}{3}.
\end{aligned}$$

Problem 4.121 The quadratic equation $2x^2 - (\sqrt{3}+1)x + m = 0$ has two roots $\sin\theta$

and $\cos\theta$, find the value of $\dfrac{\sin\theta}{1-\cot\theta}+\dfrac{\cos\theta}{1-\tan\theta}$.

Solution: $\dfrac{\sqrt{3}+1}{2}$.

By Vieta's formulas, $\sin\theta+\cos\theta=\dfrac{\sqrt{3}+1}{2}$. Thus

$$
\begin{aligned}
\frac{\sin\theta}{1-\cot\theta}+\frac{\cos\theta}{1-\tan\theta} &= \frac{\sin^2\theta}{\sin\theta-\cos\theta}+\frac{\cos^2\theta}{\cos\theta-\sin\theta}\\
&= \frac{\cos^2\theta-\sin^2\theta}{\cos\theta-\sin\theta}\\
&= \cos\theta+\sin\theta\\
&= \frac{\sqrt{3}+1}{2}.
\end{aligned}
$$

Problem 4.122 Given that $\sin\alpha-\cos\alpha=-\dfrac{\sqrt{5}}{5}$, and $180°<\alpha<270°$, find the value of $\tan\alpha$.

Solution: 2.

There are multiple ways to solve this problem. Here is one: $\sin\alpha-\cos\alpha=\sqrt{2}(\sin\alpha\cos45°-\cos\alpha\sin45°)=\sqrt{2}\sin(\alpha-45°)=-\dfrac{\sqrt{5}}{5}$, thus $\sin(\alpha-45°)=-\dfrac{\sqrt{10}}{10}$. Since the sine value is negative, $\alpha-45°$ is still in the third quadrant. Hence, $\cos(\alpha-45°)=-\sqrt{1-\sin^2(\alpha-45°)}=-\dfrac{3\sqrt{10}}{10}$, and $\tan(\alpha-45°)=\dfrac{1}{3}$.
And, $\tan(\alpha-45°)=\dfrac{\tan\alpha-1}{1+\tan\alpha}=\dfrac{1}{3}$, therefore $\tan\alpha=2$.

Problem 4.123 Given that $\sin x=\dfrac{4}{5}$, where $x\in\left(\dfrac{\pi}{2},\pi\right)$, and $\cos y=-\dfrac{5}{13}$, where $y\in\left(\pi,\dfrac{3\pi}{2}\right)$. Find the value of $\cos(x+y)$.

Solution: $\dfrac{63}{65}$.

$\cos x=-\dfrac{3}{5}$, $\sin y=-\dfrac{12}{13}$, so $\cos(x+y)=\cos x\cos y-\sin x\sin y=\dfrac{3}{13}+\dfrac{48}{65}=\dfrac{63}{65}$.

Problem 4.124 Find maximum and minimum values of

(a) $\dfrac{1}{2}\cos x-\dfrac{\sqrt{2}}{2}\sin x$.

Solution: $\pm\sqrt{3}/2$.

Let $A = \sqrt{\left(\frac{1}{2}\right)^2 + \left(\frac{\sqrt{2}}{2}\right)^2} = \frac{\sqrt{3}}{2}$, $\theta = \arcsin\left(\frac{1/2}{A}\right)$,

then $\sin\theta = \frac{1/2}{A}$ and $\cos\theta = \frac{\sqrt{2}/2}{A}$, and

$\frac{1}{2}\cos x - \frac{\sqrt{2}}{2}\sin x = A\left(\sin\theta\cos x - \cos\theta A\sin x\right) = A\sin\left(\theta - x\right).$

Since $|sin(\theta - x)| \le 1$, the maximum value of the given expression is A, and the minimum value is $-A$.

(b) $3\sin x + 4\cos x$
   Solution: ± 5.
   Let $\theta = \arccos\frac{3}{5}$, then

$$
\begin{aligned}
3\sin x + 4\cos x &= 5\left(\frac{3}{5}\sin x + \frac{4}{5}\cos x\right) \\
&= 5(\cos\theta\sin x + \sin\theta\cos x) \\
&= 5\sin(x + \theta).
\end{aligned}
$$

Therefore the maximum and minimum values are ± 5.

(c) $a\sin x + b\cos x$, where $ab \ne 0$.
   Solution: $\pm\sqrt{a^2 + b^2}$.
   Similar to the previous problems.

$$
\begin{aligned}
a\sin x + b\cos x &= \sqrt{a^2 + b^2}\left(\frac{a}{\sqrt{a^2 + b^2}}\sin x + \frac{b}{\sqrt{a^2 + b^2}}\cos x\right) \\
&= \sqrt{a^2 + b^2}\sin\left(x + \arccos\left(\frac{a}{\sqrt{a^2 + b^2}}\right)\right).
\end{aligned}
$$

Therefore the maximum and minimum values are $\pm\sqrt{a^2 + b^2}$.

(d) $\sin x + \sin\left(x + \frac{\pi}{4}\right)$
   Solution: $\pm\sqrt{2 + \sqrt{2}}$.
   $\sin x + \sin\left(x + \frac{\pi}{4}\right) = \left(1 + \frac{\sqrt{2}}{2}\right)\sin x + \frac{\sqrt{2}}{2}\cos x$, then apply the method similar to the previous problems.

Problem 4.125 Given $x, y \in \left[-\dfrac{\pi}{4}, \dfrac{\pi}{4}\right]$, and $a \in \mathbb{R}$, and

$$\begin{cases} x^3 + \sin x - 2a = 0 \\ 4y^3 + \dfrac{1}{2}\sin 2y + a = 0 \end{cases}$$

Find $\cos(x + 2y)$.

Solution: Since $f(t) = t^3 + \sin t$ is an odd function and strictly increasing in $\left[-\dfrac{\pi}{2}, \dfrac{\pi}{2}\right]$, and we are given that $f(x) = 2a$ and $f(2y) = -2a = -f(x) = f(-x)$, so $2y = -x$, therefore $x + 2y = 0$, thus $\cos(x + 2y) = 1$.

Problem 4.126 Let $0 < x < \dfrac{\pi}{4}$, arrange the following four numbers in increasing order:

$$(\tan x)^{\tan x}, \quad (\tan x)^{\cot x}, \quad (\cot x)^{\tan x}, \quad (\cot x)^{\cot x}$$

Solution: Since $0 < \tan x < 1 < \cot x$, $(\tan x)^{\cot x} < (\tan x)^{\tan x} < (\cot x)^{\tan x} < (\cot x)^{\cot x}$.

Problem 4.127 If $\sin x = -\dfrac{4}{5}$ and $\tan x < 0$, find the exact value of $\cos 3x$.

Solution: $-\dfrac{117}{125}$.

$\cos x = \dfrac{3}{5}$, so $\cos 3x = 4\cos^3 x - 3\cos x = -\dfrac{117}{125}$.

Problem 4.128 Given obtuse angles α and β, satisfying: $\sin \alpha = \dfrac{12}{13}$, $\cos(\beta - \alpha) = \dfrac{3}{5}$. Find $\sin \beta$.

Solution: $\dfrac{16}{65}$.

We don't know which angle is larger, so $\sin(\beta - \alpha) = \pm\dfrac{4}{5}$. We know that $\cos \alpha = -\dfrac{5}{13}$. Then $\sin \beta = \sin(\alpha + \beta - \alpha) = \sin \alpha \cos(\beta - \alpha) + \cos \alpha \sin(\beta - \alpha) = \dfrac{12}{13} \cdot \dfrac{3}{5} \pm \dfrac{-5}{13} \cdot \dfrac{4}{5} = \dfrac{36 \mp 20}{65}$. To find out which is the correct answer, we calculate $\cos \beta = \cos(\alpha + \beta - \alpha) = \cos \alpha \cos(\beta - \alpha) - \sin \alpha \sin(\beta - \alpha) = -\dfrac{5}{13} \cdot \dfrac{3}{5} - \dfrac{12}{13} \cdot \left(\pm\dfrac{4}{5}\right) = \dfrac{33}{65}$ or $-\dfrac{63}{65}$, for obtuse angle β, $\cos \beta < 0$, so we take $\sin(\beta - \alpha) = \dfrac{4}{5}$. Therefore the answer is $\dfrac{16}{65}$.

Problem 4.129 Find the smallest positive period of $f(x) = |\sin x| + |\cos x|$.

Solution: $\pi/2$.

$f(x) = \sqrt{1 + |\sin 2x|}$. Since $\sin 2x$ has the smallest positive period π, the smallest positive period of $|\sin 2x|$ is $\dfrac{\pi}{2}$, and the one for $f(x)$ is the same.

Problem 4.130 Simplify:

(a) $3(\sin^4 x + \cos^4 x) - 2(\sin^6 x + \cos^6 x)$
   Solution: 1.

$$
\begin{aligned}
sin^6 x + \cos^6 x &= (\sin^2 x + \cos^2 x)(\cos^4 x - \cos x \sin x + \sin^4 x) \\
&= \cos^4 x - \cos x \sin x + \sin^4 x,
\end{aligned}
$$

so

$$
\begin{aligned}
&3(\sin^4 x + \cos^4 x) - 2(\sin^6 x + \cos^6 x) \\
=\; &3(\sin^4 x + \cos^4 x) - 2(\cos^4 x - \cos x \sin x + \sin^4 x) \\
=\; &\sin^4 x + 2 \sin x \cos x + \cos^4 x \\
=\; &1.
\end{aligned}
$$

(b) $\sqrt{\sin^4 x + 4\cos^2 x} - \sqrt{\cos^4 x + 4\sin^2 x}$
   Solution: $\cos 2x$.

$$
\begin{aligned}
&\sqrt{\sin^4 x + 4\cos^2 x} - \sqrt{\cos^4 x + 4\sin^2 x} \\
=\; &\sqrt{\sin^4 x - 4\sin^2 x + 4} - \sqrt{\cos^4 x - 4\cos^2 x + 4} \\
=\; &\sqrt{(\sin^2 x - 2)} - \sqrt{(\cos^2 x - 2)} \\
=\; &(2 - \sin^2 x) - (2 - \cos^2 x) \\
=\; &\cos^2 x - \sin^2 x \\
=\; &\cos 2x
\end{aligned}
$$

(c) $\cos x + \cos 2x + \cos 3x + \cdots + \cos nx$
   Solution: $\dfrac{\cos \dfrac{(n+1)x}{2} \sin \dfrac{nx}{2}}{\sin \dfrac{x}{2}}$.

Let $C = \cos x + \cos 2x + \cos 3x + \cdots + \cos nx$, then multiply $\sin\dfrac{x}{2}$ and telescope:

$$
\begin{aligned}
\left(\sin\frac{x}{2}\right)C &= \sum_{k=1}^{n}\sin\frac{x}{2}\cos kx \\
&= \sum_{k=1}^{n}\frac{1}{2}\left(\sin\left(k+\frac{1}{2}\right)x - \sin\left(k-\frac{1}{2}\right)x\right) \\
&= \frac{1}{2}\left(\sin\frac{2n+1}{2}x - \sin\frac{1}{2}x\right) \\
&= \cos\frac{(n+1)x}{2}\sin\frac{nx}{2},
\end{aligned}
$$

therefore, $C = \dfrac{\cos\dfrac{(n+1)x}{2}\sin\dfrac{nx}{2}}{\sin\dfrac{x}{2}}$.

(d) $\displaystyle\sum_{k=0}^{n}\arctan\frac{1}{k^2+k+1}$

Solution: $\arctan(n+1)$.

$\arctan\dfrac{1}{k^2+k+1} = \arctan(k+1) - \arctan k$, and telescope.

(e) $\displaystyle\sum_{k=1}^{n}\arctan\frac{1}{2k^2}$

Solution: $\arctan(2n+1) - \dfrac{\pi}{4}$.

$\arctan\dfrac{1}{2k^2} = \arctan(2k+1) - \arctan(2k-1)$, so

$$\sum_{k=1}^{n}\arctan\frac{1}{2k^2} = \arctan(2n+1) - \arctan 1 = \arctan(2n+1) - \frac{\pi}{4}.$$

Note: this answer also equals $\arctan\dfrac{n}{n+1}$.

Problem 4.131 Let x and y be real numbers such that

$$\sin x + \sin y = \frac{\sqrt{2}}{2} \quad \text{and} \quad \cos x + \cos y = \frac{\sqrt{6}}{2},$$

find the value of $\sin(x+y)$.

Solution: $\dfrac{\sqrt{3}}{2}$.

Square both equations and add, $2 + \cos(x-y) = 2$, so $\cos(x-y) = 0$. Multiply the equations, $\dfrac{1}{2}\sin 2x + \dfrac{1}{2}\sin 2y + \sin(x+y) = \dfrac{\sqrt{3}}{2}$.

Also by the Sum-to-Product formula, $\frac{1}{2}\sin 2x + \frac{1}{2}\sin 2y = \sin(x+y)\cos(x-y) = 0$, so $\sin(x+y) = \frac{\sqrt{3}}{2}$.

Problem 4.132 Solve the following inequality for $0 \le x \le 2\pi$:

$$1 + \tan x > 2\cos x + 2\sin x$$

Solution: $\left(\frac{\pi}{3}, \frac{\pi}{2}\right) \cup \left(\frac{3\pi}{4}, \frac{3\pi}{2}\right) \cup \left(\frac{5\pi}{3}, \frac{7\pi}{4}\right)$.

Note that $2\cos x + 2\sin x = 2\cos x(1 + \tan x)$, thus

$$\begin{aligned}
1 + \tan x - 2\cos x + 2\sin x &> 0, \\
(1 + \tan x) - 2\cos x(1 + \tan x) &> 0, \\
(1 + \tan x)(1 - 2\cos x) &> 0.
\end{aligned}$$

There are two possibilities: (1) $1 + \tan x > 0$ and $1 - 2\cos x > 0$; (2) $1 + \tan x < 0$ and $1 - 2\cos x < 0$.

Case (1): $\tan x > -1$ and $\cos x < \frac{1}{2}$. For $\tan x > -1$, we have $x \in \left[0, \frac{\pi}{2}\right) \cup \left(\frac{3\pi}{4}, \frac{3\pi}{2}\right) \cup \left(\frac{7\pi}{4}, 2\pi\right]$; for $\cos x < \frac{1}{2}$, $x \in \left(\frac{\pi}{3}, \frac{5\pi}{3}\right)$. The intersection of these two sets is $\left(\frac{\pi}{3}, \frac{\pi}{2}\right) \cup \left(\frac{3\pi}{4}, \frac{3\pi}{2}\right)$.

Case (2): $\tan x < -1$ and $\cos x > \frac{1}{2}$. For $\tan x < -1$, we have $x \in \left(\frac{\pi}{2}, \frac{3\pi}{4}\right) \cup \left(\frac{3\pi}{2}, \frac{7\pi}{4}\right)$; for $\cos x > \frac{1}{2}$, $x \in \left[0, \frac{\pi}{3}\right) \cup \left(\frac{5\pi}{3}\right]$. The intersection of these two sets is $\left(\frac{5\pi}{3}, \frac{7\pi}{4}\right)$.

Therefore the final answer is the union of the results from both cases: $\left(\frac{\pi}{3}, \frac{\pi}{2}\right) \cup \left(\frac{3\pi}{4}, \frac{3\pi}{2}\right) \cup \left(\frac{5\pi}{3}, \frac{7\pi}{4}\right)$.

Problem 4.133 Solve the following equations:
  (a) $\sqrt{3}\sin x = \cos x$
  Solution: $\pi/6 + k\pi$, $k \in \mathbb{Z}$.
  The equation becomes $\tan x = \frac{1}{\sqrt{3}}$, thus the solution between 0 and π is $\frac{\pi}{6}$, and since the period of $\tan x$ is π, the full solution is $\pi/6 + k\pi$, k is any integer.

(b) $3\cos^2 x = \sin x \sin 2x$

Solution: $k\pi + \dfrac{\pi}{2}$ or $2k\pi \pm \dfrac{\pi}{3}$, $k \in \mathbb{Z}$.

The equation becomes

$$
\begin{aligned}
3\cos^2 x &= \sin x \cdot 2\sin x \cos x, \\
3\cos^2 x &= 2\sin^2 x \cos x, \\
3\cos^2 x &= 2(1 - \cos^2 x)\cos x, \\
2\cos^3 x + 3\cos^2 x - 2\cos x &= 0, \\
\cos x (2\cos x - 1)(\cos x + 2) &= 0.
\end{aligned}
$$

Thus $\cos x = 0$ or $\cos x = \dfrac{1}{2}$. Final answer: $k\pi + \dfrac{\pi}{2}$ or $2k\pi \pm \dfrac{\pi}{3}$, $k \in \mathbb{Z}$.

(c) $\tan x - 3\cot x = 0$, for $x \in [0, 2\pi]$.

Solution: $\pi/3, 2\pi/3, 4\pi/3, 5\pi/3$.

The equation becomes $\tan^2 x = 3$, thus $\tan x = \pm\sqrt{3}$. There are four solutions in $[0, 2\pi]$: $\dfrac{\pi}{3}, \dfrac{2\pi}{3}, \dfrac{4\pi}{3}, \dfrac{5\pi}{3}$.

(d) $\sin^2 x + 2\sin x \cos x = 3\cos^2 x$

Solution: $k\pi - \arctan 3$ or $k\pi + \pi/4$, $k \in \mathbb{Z}$.

The equation can be factored:

$$
\begin{aligned}
\sin^2 x + 2\sin x \cos x - 3\cos^2 x &= 0, \\
(\sin x - \cos x)(\sin x + 3\cos x) &= 0.
\end{aligned}
$$

So either $\sin x - \cos x = 0$, which means $\tan x = 1$, so $x = k\pi + \dfrac{\pi}{4}$ ($k \in \mathbb{Z}$); or $\sin x + 3\cos x = 0$, which means $\tan x = -3$, thus $x = k\pi - \arctan 3$ ($k \in \mathbb{Z}$).

5. Combinatorics

Problem 5.1 How many factors of 2^{95} are larger than $1,000,000$?

<u>Solution</u>: 76.

Note 2^{95} has 96 factors: $2^0, 2^1, \ldots, 2^{95}$. Further, $2^{10} = 1024$ so $2^{20} = 1048576$. Thus, $2^{19} = 524288 < 1000000 < 1048576 = 2^{20}$. Hence, 20 of the factors are less than 1000000, so $96 - 20 = 76$ are greater.

Problem 5.2 Telephone numbers in *Land of Nosix* have 7 digits, and the only digits available are $\{0, 1, 2, 3, 4, 5, 7, 8\}$. No telephone number may begin in $0, 1,$ or 5. Find the number of telephone numbers possible that meet the following criteria:

(a) you may have repeated digits.

<u>Solution</u>: $5 \cdot 8^6 = 1310720$.

Pick each digit one at a time, remembering only 5 of the 8 digits can be used as the beginning of the phone number.

(b) you may not have repeated digits.

<u>Solution</u>: $5 \cdot 7 \cdot 6 \cdot 5 \cdot 4 \cdot 3 \cdot 2 = 5 \cdot \dfrac{7!}{1!} = 25200$.

Pick each digit one at a time, remembering that since we are not allowed to repeat digits, one less is available for each digit.

(c) you may have repeated digits, but the phone number must be even.

<u>Solution</u>: $5 \cdot 8^5 \cdot 4 = 655360$.

Proceed similarly to part (a), but the last digit must be even.

(d) you may not have repeated digits, and the phone number must be odd.

Solution: $5 \cdot 6 \cdot 5 \cdot 4 \cdot 3 \cdot 2 \cdot 2 + 4 \cdot 6 \cdot 5 \cdot 4 \cdot 3 \cdot 2 \cdot 2 = 5 \cdot \frac{6!}{1!} \cdot 2 + 4 \cdot \frac{6!}{1!} \cdot 2 = 12960$.

We now pick the *last* digit first, which must be chosen from $1,3,5,7,9$. We deal with two cases: (i) the last digit is 1 or 5, and (ii) otherwise. We can then proceed as in part (b).

Problem 5.3 Suppose you have a group of 12 people. How many different photographs are there of everyone lined up if:

(a) all the people look different?
Solution: $12! = 479001600$.
This is a permutation $_{12}P_{12}$.

(b) 4 of the people are identical quadruplets who have dressed identically?
Solution: $\frac{12!}{4!} = \binom{12}{4} \cdot 8! = 19958400$.

Two methods are presented above. For the first, we divide the answer from (a) by 2 because we do not care about the order of the quadruplets (since we cannot tell them apart). For the second, we first choose 4 spots for the quadruplets to stand, and then fill in the other 8 people around them.

(c) 3 of the people are a family and must stand next to each other?
Solution: $3! \cdot 10! = 21772800$.
There are $6 = 3!$ arrangements of the family. If we then treat them as a single 'object' we must arrange 10 'objects', which is a permutation.

(d) 3 of the people do not get along, and cannot all 3 be right next to each other in a group?
Solution: $12! - 3! \cdot 10! = 457228800$.
We count the total number of arrangement and subtract off the number of arrangements where they are next to each other (which is equivalent to what we counted in part (c)).

Problem 5.4 Suppose you have a student group with 15 males and 10 females.

(a) How many ways are there to pick a group of 5 males and 5 females?
Solution: $\binom{15}{5} \cdot \binom{10}{5} = 756756$.
Pick the groups of males and females separately. For each we have a combination.

(b) How many ways are there to pick an Executive Committee of 5 members and a Party Planning Committee of 5 members? Members can be on both committees at once, but each committee must have at least one male and at least one female.

Solution: $\left(\binom{25}{5} - \binom{15}{5} - \binom{10}{5} \right)^2 = 2487515625.$

First note that since members can be on both committees at once, there are the same number of ways to choose either committee. With no restrictions, there are $\binom{25}{5}$ ways to choose a committee of 5 members from 25 total. We use complementary counting to subtract the number of committees made of only males or only females.

(c) Suppose you still need to pick an Executive Committee and Party Planning Committee (each with 5 members). This time, only the Executive Committee is required to have a member of each gender, but now members are *not* allowed to be on both committees at once.

Solution: $\left(\binom{25}{5} - \binom{15}{5} - \binom{10}{5} \right) \cdot \binom{20}{5} = 773262000.$

Picking the Executive Committee is the same as in part (b), and now the Party Planning Committee is chosen without restrictions from the remaining 20 members.

Problem 5.5 The number 3 can be expressed as a sum of one or more positive integers in four ways, namely, as 3, $1+2$, $2+1$, and $1+1+1$.

(a) How many ways can the number 6 be expressed as the sum of one or more positive integers less than or equal to 2?

Solution: $\binom{6}{0} + \binom{5}{1} + \binom{4}{2} + \binom{3}{3} = 1 + 5 + 6 + 1 = 13.$

The integers must all be 1 or 2. We break into 4 cases based on how many 2's are used (either $0, 1, 2, 3$). In these cases (respectively) we know there are a $6, 5, 4, 3$ total integers used, so we pick which of those integers will be 2.

(b) How many total ways can 6 be expressed as the sum of one or more positive integers?

Solution: 32.

Note that there are many more cases to deal with here (try to work them out yourself!). We use a trick: $6 = 1 + 1 + 1 + 1 + 1 + 1$, and we can consider the "1"s as 'sticks", and if several "1"s are together we treat them as the number of sticks in the same pile. At each gap between two "1"s, the plus sign can be either present or absent, so there are 2 choices for each gap, and there are 5 gaps. Therefore there are $2^5 = 32$ possible sums.

Problem 5.6 A bookshelf contains 5 German books, 7 Spanish books, and 8 French books. Each book is different from one another.

(a) How many different arrangements can be done of these books?

Solution: 20!.
This is a permutation.

(b) How many different arrangements can be done if books of each language must be next to each other?
Solution: $(5! \cdot 7! \cdot 8!) \cdot 3!$.
There are $5!, 7!, 8!$ ways (respectively) to order the German, Spanish, and French books. Then treat each group as a single 'object' and order the 3 languages (3! ways).

(c) How many different arrangements can be done if no two German books must be next to each other?
Solution: $15! \cdot (16 \cdot 15 \cdot 14 \cdot 13 \cdot 12) = 15! \cdot \dfrac{16!}{11!}$.
First arrange the 15 Spanish and French books. Putting these books creates $15 + 1 = 16$ 'spaces' (before the first book, after the last book, and in between the others). Since the German books cannot be next to each other, the four German books must be put in these spaces (at most one book per space).

Problem 5.7 Suppose you have 40 identical balls and 8 numbered boxes. How many ways are there to put the balls into the boxes if:

(a) there are no restrictions?
Solution: $\dbinom{40+8-1}{40} = 62891499$.
This is a routine example of the non-negative version of stars and bars.

(b) each box has at least two balls?
Solution: $\dbinom{24+8-1}{24} = 2629575$.
First put two balls in each box. Since the balls are identical, there is only 1 way to do this. We are left with 24 balls and 8 boxes, and we use the non-negative version of stars and bars again.

(c) no box has more than 5 balls?
Solution: 1.
Note the only possibility here is that each box has exactly 5 balls.

(d) the first box has exactly 10 balls?
Solution: $\dbinom{30+7-1}{30} = 1947792$.
First put 10 balls into the first box. We are left with 30 balls and $8 - 1 = 7$ boxes (since the first box cannot have any more balls).

Problem 5.8 Suppose you have 10 blue, 10 red, and 10 green balls. You want to arrange the balls so that no two green balls are next to each other. How many ways are there to do this if

(a) the balls are in a row and each ball is numbered?

Solution: $20! \cdot \dfrac{21!}{11!}$.

Arrange the blue and red balls first. This is a permutation of the 20 total blue and red balls. This creates $20+1$ spaces for the green balls. We then arrange the green balls in those spaces (here choosing which space to put each of the 10 green balls in).

(b) the balls are in a row and each ball is identical?

Solution: $\dbinom{20}{10} \cdot \dbinom{21}{10}$,

Same strategy as part (a), except since the balls are identical we use combinations rather than permutations when deciding where to place the blue/red and green balls.

Second Solution: We divide the answer in part (a) by $(10!)^3$, as we do not care about how the balls are numbered. Double check this leads to the same answer!

(c) the balls are in a circle and each ball is numbered?

Solution: $\dfrac{20!}{20} \cdot \dfrac{20!}{10!}$.

Arrange the blue and red balls first (this is a circular permutation). Since they are arranged in a circle, this creates 20 spaces for the green balls (instead of the 21 it creates in a line).

Problem 5.9 Given positive integers $1, 2, 3, \ldots, n$. Let a permutation of these numbers satisfy the requirement that, for each number, it is either greater than all the numbers before it, or less than all the numbers before it. How many such permutations are there?
Solution: 2^{n-1}.
Start with the last number: it is either 1 or n (2 choices). For the second number, it is either the largest remaining, or the smallest remaining (2 choices again). This pattern continues for all the numbers except the first (as there is only one number remaining).

Problem 5.10 Consider the number 100000.

(a) How many factors does it have?
Solution: 36.
$100000 = 2^5 \cdot 5^5$ so it has $(5+1) \cdot (5+1) = 36$ factors.

(b) How many ways are there to represent it as the product of 2 factors if we consider products that differ in the order of factors to be different?
Solution: 36.
Note this is the same as part (a) as each factor comes in a pair.

Second Solution: Alternatively, $100000 = 2^5 \cdot 5^5 = a \cdot b = (2^m 5^n)(2^p 5^q)$, so it is actually stars and bars for both the powers of 2 and powers of 5: $m + p = 5$ and $n + q = 5$, zeros are allowed. Therefore the answer is $\binom{5+2-1}{5}^2 = 36$.

(c) How many ways are there to represent it as the product of 3 factors if we consider products that differ in the order of factors to be different?
Solution: 441.
$100000 = 2^5 \cdot 5^5 = a \cdot b \cdot c = (2^m 5^n)(2^p 5^q)(2^r 5^s)$, so we need $m + p + r = 5$ and $n + q + s = 5$, zeros are allowed. Using stars and bars, the answer is $\binom{5+3-1}{5}^2 = 441$.

Problem 5.11 Suppose you have 8 red cards and 25 black cards. Assume all the cards of the same color are identical. Deal the cards out in a line.

(a) How many different arrangements of the cards are there?
Solution: $\binom{33}{8} = 13884156$.
Since cards of the same color are identical, decide which 8 of the 33 slots are for the red cards.

(b) Repeat part (a), if there must be at least 2 black cards between all the red cards.
Solution: $\binom{11+9-1}{11} = 75582$.
Arrange the red cards (only 1 way). Then place 2 black cards in between the cards (only 1 way) using $2 \cdot 7 = 14$ black cards (we do not *need* black cards at the end). Then the remaining 11 cards can arranged in the 9 spaces created by the red cards (now including the endpoints) using stars and bars.

Problem 5.12 Suppose you place 9 rings on the 3 mid fingers of your left hand (that is, not on your thumb or your pinky). How many different outcomes are possible if

(a) all the rings are identical, and no finger has more than 3 rings?
Solution: 1.
There must be exactly 3 rings on each finger.

(b) all the rings are different, and no finger has more than 3 rings?
Solution: 9!.
Note once the rings are placed on the fingers, we have an order on all the rings, so this is equivalent to ordering the rings in a line.
Second Solution: We first choose which rings go on each finger (there are $\binom{9}{3}\binom{6}{3}\binom{3}{3}$ ways to do this). Then we must arrange the three rings on each

finger (3! for each finger). Double check that $\binom{9}{3}\binom{6}{3}\binom{3}{3}\cdot(3!)^3 = 9!$.

(c) all the rings are identical, and no finger has more than 8 rings?

Solution: $\binom{9+3-1}{9} - 3 = 52$.

Note that the opposite of no finger has more than 8 rings is that one of the fingers has all 9 rings. Hence, use stars and bars to calculate the total number of outcomes and subtract the 3 ways where all the rings are on a single finger.

(d) all the rings are different, and all the rings are on a single finger?

Solution: $3 \cdot 9! = 1088640$.

There are 9! ways to order the rings, which can then be placed on one of the 3 fingers.

Problem 5.13 How many ways are there to write 200 as the sum of three non-negative integers (we care about the order of the numbers)

(a) in total?

Solution: $\binom{200+3-1}{200} = 20301$.

This is stars and bars.

(b) if all three numbers must be different?

Solution: $\binom{200+3-1}{200} - 303 = 19998$.

We use complementary counting. Note it is impossible for all 3 numbers to be the same. What if we have a pair of repeated numbers? That is we have $a+a+b = 2a+b = 200$ for non-negative a,b. We can list the outcomes as ordered pairs (a,b): $(0,200),(1,198),(2,196),\ldots,(100,0)$, that is, there are 101 possibilities. However, each of these can be arranged in 3 different ways ($a+a+b = a+b+a = b+a+a$), for a total of 303 ordered outcomes. We subtract this from the total number of outcomes in (a).

Problem 5.14 What is the coefficient of x^2 in $(x+3)^8$?

Solution: $\binom{8}{6} \cdot 3^6 = 20412$.

Use the Binomial Theorem with $a = x$ and $b = 3$.

Problem 5.15 Practice on Pascal's Triangles.

(a) Show that $\binom{8}{0} + \binom{8}{2} + \binom{8}{4} + \binom{8}{6} + \binom{8}{8} = \binom{8}{1} + \binom{8}{3} + \binom{8}{5} + \binom{8}{7}$.

Solution: Move all the terms to one side and use the Binomial Theorem ($a = 1, b = -1$ so $(1-1)^8 = 0$).

(b) Compute $2\binom{8}{0} + 2\binom{8}{2} + 2\binom{8}{4} + 2\binom{8}{6} + 2\binom{8}{8}$.

Solution: $2^8 = 256$.

Use part (a) to show the above is equal to $\binom{8}{0} + \binom{8}{1} + \cdots + \binom{8}{8} = 2^8$ using the Binomial Theorem ($a = 1, b = 1$ so $(1+1)^8 = 2^8$).

Problem 5.16 Find the constant term in the expansion of $\left(\sqrt{x} + \dfrac{1}{\sqrt{x}} - 2\right)^5$.

Solution: $(-1)^5 \binom{10}{5} = -252$.

Let $y^2 = \sqrt{x}$, then the expression becomes

$$\left(y^2 + \frac{1}{y^2} - 2\right)^5 = \left(y - \frac{1}{y}\right)^{10},$$

so use the Binomial Theorem to find the coefficient of the constant term, where the constant term is $y^5 \cdot \left(\dfrac{1}{y}\right)^5 = 1$.

Problem 5.17 Simplify the following

(a) $\displaystyle\sum_{k=1}^{n} k\binom{n}{k}$.

Solution: $n2^{n-1}$.

Use reduction and then the binomial theorem for $(1+1)^{n-1} = 2^{n-1}$: $\displaystyle\sum_{k=1}^{n} k\binom{n}{k} =$

$$\sum_{k=1}^{n} n\binom{n-1}{k-1} = n \cdot \sum_{k=0}^{n-1}\binom{n-1}{k} = n2^{n-1}.$$

(b) $\displaystyle\sum_{k=0}^{n} \frac{1}{k+1}\binom{n}{k}$.

Solution: $\dfrac{2^{n+1} - 1}{n+1}$.

Use reduction, $\binom{n}{k} = \dfrac{k+1}{n+1}\binom{n+1}{k+1}$, shift the index ($l = k+1$), and use the binomial theorem:

$$\sum_{k=0}^{n} \frac{1}{k+1} \binom{n}{k} = \sum_{k=0}^{n} \frac{1}{k+1} \frac{k+1}{n+1} \binom{n+1}{k+1}$$

$$= \sum_{k=0}^{n} \frac{1}{n+1} \binom{n+1}{k+1}$$

$$= \frac{1}{n+1} \sum_{k=0}^{n} \binom{n+1}{k+1}$$

$$= \frac{1}{n+1} \left[\sum_{l=0}^{n+1} \binom{n+1}{l} - 1 \right]$$

$$= \frac{2^{n+1} - 1}{n+1}$$

Problem 5.18 Suppose 8 dinner guests attend a dinner party and are seated at a circular table. Two of the guests are a couple that must sit together. If one of the seats is the "Head of the Table", how many ways are there to seat the guests? Hint: Be careful!

Solution: $2! \cdot \frac{7!}{7} \cdot 8 = 11520$.

There are $2! \cdot \frac{7!}{7}$ ways to seat the guests around without the "Head of the Table" seat. We then have to choose which guest sits in that seat, for which there are 8 choices.

Problem 5.19 How many numbers between $1 - 100$ are a multiple of either $5, 7, 9$, or 11?

Solution: $20 + 14 + 11 + 9 - 2 - 2 - 1 - 1 - 1 - 1 = 46$.

Use PIE for A, B, C, D (multiples of $5, 7, 9, 11$) and note that the intersection of any 3 or more sets is empty. For example, $n(B) = \lfloor 100/7 \rfloor = 14$ and $B \cap C$ is all multiples of $\gcd(7,9) = 63$ so $n(B \cap C) = \lfloor 100/63 \rfloor = 1$.

Problem 5.20 Suppose you want to bring a collection of 12 sodas to a party. You can choose from 6 types (and all that matters is how many of each soda you bring). How many ways can you do this if

   (a) there are no restrictions on the types of soda you bring?

     Solution: $\binom{12+6-1}{12} = 6188$.

   Routine non-negative stars and bars.

(b) one of your friends at the party really likes Fanta (one of the 6 types) so you want to bring at least 6 Fantas to the party?

Solution: $\binom{6+6-1}{6} = 462$.

After we set aside 6 Fantas, we still have 6 sodas which can be any of the 6 types.

Problem 5.21 Suppose you have 12 identical balls and 4 numbered boxes. How many ways are there to put the balls in the boxes if the first box has at least 3 balls, but no more than 5 balls?

Solution: $\binom{9+3-1}{9} + \binom{8+3-1}{8} + \binom{7+3-1}{7} = 136$.

We break into cases based on whether the first box has $3, 4$, or 5 balls. In every case, we distribute the remaining balls among the 2nd, 3rd, and 4th boxes using non-negative stars and bars.

Problem 5.22 Suppose you have a sequence of 10 integers, and the first integer is 10. Every integer after the first is either one larger or smaller than the previous.

(a) How many different sequences of integers are possible?
Solution: 2^9.
There are 2 choices for each integer after the first (either $+1$ or -1 from the previous).

(b) How many different possibilities are there for the last integer in the sequence?
Solution: 11.
Every even number between 0 and 20 is possible $(0, 2, 4, \ldots, 18, 20)$.

Problem 5.23 In how many ways can a necklace be made using 5 identical red beads and 2 identical blue beads? Hint: Try brute force.
Solution: 3.
Note that the beads can be rotated or reflected and still produce the same necklace. The two blue beads are either together, or separated by 1 or 2 red beads, so there are 3 possible necklaces.

Problem 5.24 Suppose you have 20 people and need to form 3 (non-empty) committees. The number of people on each committee must be a multiple of 5. If all three committees are different (Executive, Party Planning, etc.), how many ways can the committees be chosen?

Solution: $3 \cdot \dfrac{20!}{10! \cdot (5!)^2} = 3 \cdot \binom{20}{10} \cdot \binom{10}{5} \cdot \binom{5}{5} = 139675536$.

Note that one committee must have 10 members and the other committees 5. There are 3 ways to decide which committee has 10 members. After we decide which committee has 10 members, we need to divide the 30 people into 3 groups, of size $10, 5, 5$.

Problem 5.25 Suppose 10 girls and 20 boys sit around a table. The girls are grouped in 5 pairs (which sit together), and in between each pair is at least 2 boys. How many arrangements are there? Hint: Work slowly, and break the problem into pieces.

Solution: $(2!)^5 \cdot \dfrac{5!}{5} \cdot \dbinom{10+5-1}{10} \cdot 20! = 187033721102193917952000.$

First arrange each pair of girls (2! ways each), and then the 5 pairs in a circular permutation (a circular permutation: $5!/5$). This creates 5 spaces. Then decide how the boys will fit in these spaces, assuming for now they are all identical (first putting 2 in each space, then using stars and bars (so left with 10 boys and 5 spaces so $\dbinom{10+5-1}{10}$ different ways). Finally, arrange the 20 boys (a permutation 20!).

Problem 5.26 Suppose you have 8 different books. You want to use the books as gifts for 3 of you friends. How many ways are there to give out the gifts? Be nice: each friend gets at least one book! (Every gift is determined only by which books are given.)

Solution: $3^8 - 3 \cdot 2^8 + 3 = 5796.$

There are 3^8 ways to give out the books with no restrictions. Then use PIE to subtract off the cases where not everyone gets a book. Let A, B, C be the events that the first, second, third friend respectively do not get a book. We have $n(A) = n(B) = n(C) = 2^8$, $n(A \cap B) = n(A \cap C) = n(B \cap C) = 1$, and $n(A \cap B \cap C) = 0$.

Problem 5.27 A train with 15 passengers must make 15 stops.

   (a) How many ways are there for the passengers to get off the train at the stops, if not all the passengers get off at the same stop?

      Solution: $15^{15} - 15 = 437893890380859360.$

      Each passenger has 15 choices for when to get off, but there are 15 ways in which all the passengers get off at the same stop.

   (b) Repeat part (a) if we only care about the number of passengers getting off at each stop?

      Solution: $\dbinom{15+15-1}{15} - 15 = 77558745.$

      Now the passengers are considered identical so we use stars and bars. There are still 15 ways they all get off at the same stop.

Problem 5.28 Suppose 5 people get in an elevator on Floor 0. The people leave the elevator somewhere between (inclusive) Floor 1 and Floor 10.

   (a) If we only care about how many people get of at each floor, how many ways can the people get off?

Solution: $\binom{5+10-1}{5} = 2002.$

This is stars and bars.

(b) Suppose the 5 people all get off at different floors. If we only care about what collection of floors the elevator stops on, how many different collections are there?

Solution: $\binom{10}{5}.$

We simply choose which 5 of the 10 floors the elevator stops on, a combination since we only care about the collection of floors.

Problem 5.29 Suppose we have four black, four white, and four green balls. How many ways are there to put the 12 balls into 6 distinguishable boxes if every color is put in at least 2 boxes?

Solution: $\left(\binom{4+6-1}{4} - 6\right)^3 = 1728000.$

Since we have 6 boxes, there are 6 ways for each color *not* to be put in at least two boxes. Hence, using stars and bars and complementary counting, there are $\binom{4+6-1}{4} - 6$ ways to choose how to place each color.

Problem 5.30 How many different ways are there to represent 81000 as the product of 3 factors if we consider products that differ in the order of factors to be different?

Solution: $\binom{3+3-1}{3}^2 \binom{4+3-1}{4} = 1500.$

We have $81000 = 2^3 \cdot 3^4 \cdot 5^3$. Thus, we need to solve $a+b+c = 3$ twice (for powers of 2, 5) and $a+b+c = 4$ once (for powers of 3). We use non-negative stars and bars for each.

Problem 5.31 Suppose an ant starts at the origin $(0,0)$. Ever step it takes is either $(1,1)$ or $(1,-1)$ (so it moves diagonally up and diagonally down).

(a) How many different ways can the ant move from the origin to $(20,0)$?

Solution: $\binom{20}{10} = 184756.$

Note that it must make 10 moves up (i.e. $(1,1)$) and 10 moves down (i.e. $(1,-1)$).

(b) Repeat part (a) if the ant makes a stop at $(10,0)$ along the way.

Solution: $\binom{10}{5}^2 = 63504.$

To move to $(10,0)$ it must make 5 moves up and 5 moves down. Then it repeats this method to get to $(20,0)$.

Problem 5.32 Suppose you have 8 numbered red cards and 20 identical black cards. How many ways are there to arrange the cards so that there is at least two black cards between each red card.

Solution: $8! \cdot \binom{6+9-1}{6} = 121080960$.

First arrange the 8 red cards (8! ways). Then place 2 black cards in between each of the red cards (7 spaces) and then use stars and bars to place the remaining 6 black cards (in 9 spaces in between and on the ends of the red cards).

Problem 5.33 Suppose you have 6 numbered red cards and 20 numbered black cards. How many ways are there to arrange the cards in a circle if there must be at least 2 black cards between each red card?

Solution: $\dfrac{6!}{6} \cdot \binom{8+6-1}{8} \cdot 20! = 375737386142800281600000$.

There are $\dfrac{6!}{6} \cdot \binom{8+6-1}{8}$ ways to arrange the cards if use identical placeholders for the black cards (using a circular permutation and then stars and bars, as after we put 2 black cards in each of the 6 spaces created by the red cards we have 8 black cards and 6 spaces left). We then multiply by 20!, which is the number of ways to assign the numbered black cards to the placeholders.

Problem 5.34 Suppose you have the numbers $\{0,1,2,3,4,5\}$. How many 6-digit numbers can be formed with 1 next to 0 or 2? (Use each number exactly once.) Caution: 012345 is *not* a 6 digit number so be careful!

Solution: $2! \cdot 4 \cdot 4! + 5! + 4 \cdot 4! - (4! + 3 \cdot 3!) = 366$.

Let A, B be the events that 1 is adjacent to $2, 0$ respectively. Then $n(A) = 2! \cdot 4 \cdot 4!$ if we think of $1, 2$ as a pair, as there are 4 choices for were the 0 goes. For $n(B)$ consider the groupings 10 and 01 separately to get $n(B) = 5! + 4 \cdot 4!$ (for 01, the pair 01 cannot be first, for 10 the digit 0 is automatically not first). For $A \cap B$, group $2, 1, 0$ with 1 in the middle, again considering 210 and 012 separately, so $n(A \cap B) = 4! + 3 \cdot 3!$.

Problem 5.35 Suppose 4 people Albert, Bill, Charles, and Drew run a race. It was predicted that Albert would finish first, Bill second, and Charles third. How many outcomes of the race are there where all 3 of these predictions are wrong? (Note: We are *not* making any predictions about Drew.)

Solution: 11.

We use PIE to calculate the number of ways that at least one person finishes in their predicted place, and subtract it from the $4! = 24$ total outcomes. Let A, B, C be be the event Albert, Bill, Charles (respectively) finish as predicted. Then $n(A) = n(B) = n(C) = 3!$ as we simply arrange the 3 other people. Similarly, $n(A \cap B) = n(A \cap C) =$

$n(B \cap C) = 2!$ and $n(A \cap B \cap C) = 1$. Then, using PIE we have

$$n(A \cup B \cup C) = 3 \cdot 3! - 3 \cdot 2! + 1! = 18 - 6 + 1 = 13.$$

Thus the final answer is $24 - 13 = 11$.

Problem 5.36 Suppose you give out 5 distinct books to 3 of your friends. Each friend gets at least one book. How many ways can you give out the books? Do the calculation using

(a) PIE.
   Solution: $3^5 - 3 * 2^5 + 3 = 150$.
   Let A, B, C be respectively events that the three friends get 0 books. Then $n(A) = n(B) = n(C) = 2^5$, and $n(A \cap B) = n(A \cap C) = n(B \cap C) = 1^5$ and $n(A \cap B \cap C) = 0$. Hence $n(A \cup B \cup C) = 3 \cdot 2^5 - 3$. We then subtract this from the total number of outcomes (3^5).

(b) directly using cases.
   Solution: $\binom{3}{2} \cdot \frac{5!}{1! \cdot 1! \cdot 3!} + \binom{3}{2} \cdot \frac{5!}{2! \cdot 2! \cdot 1!} = 150$.
   We can either give two friends 1 book and the third 3 books, or two friends 2 books and the third 1 book. In both cases we need to first choose which two friends (so $\binom{3}{2}$ ways in each case) and then choose which books to give to each friend (here there are 5! ways to order the books, but if we give a friend multiple books, the order doesn't matter, so we divide by 1!1!3! and 2!2!1! respectively).

Problem 5.37 Suppose a pizza place has 10 toppings available. You want to order 2 different 3-topping pizzas. Suppose repeated toppings are not allowed on a single pizza, and the order of the toppings on a pizza does not matter. If you only care which two pizzas you get, how many ways are there to make the order?
Solution: $\binom{\binom{10}{3}}{2} = \binom{120}{2} = 7140$.

Since toppings cannot be repeated, there are $\binom{10}{3} = 120$ different 2-topping pizzas. We then need to choose 2 (again without order) for our order.

Problem 5.38 How many ways are there to put 8 balls in 4 numbered boxes, so that each box gets at least one ball? Hint: It is probably easiest to use PIE and Complementary Counting.
Solution: $4^8 - 4 \cdot 3^8 + 6 \cdot 2^8 - 4 = 40824$.
Let A, B, C, D be respectively events that the four boxes get 0 balls. Then $n(A) = n(B) =$

$n(C) = n(D) = 3^8$ (note there are $\binom{4}{1} = 4$ possibilities here), $n(A \cap B) = n(A \cap C) = n(A \cap D) = \cdots = 2^8$ (note there are $\binom{4}{2} = 6$ possibilities here), and $n(A \cap B \cap C) = \cdots = 1^8$ (note there are $\binom{4}{3} = 4$ possibilities here). Lastly, $n(A \cap B \cap C \cap D) = 0$. Then use PIE and subtract from the total number of outcomes (which is 4^8).

Problem 5.39 Suppose you place 4 different rings on the 3 mid fingers of your left hand. How many different outcomes are possible? Hint: Cases!

Solution: $3 \cdot \binom{4}{4} \cdot 4! + (3 \cdot 2) \cdot \binom{4}{3} \cdot (3! \cdot 1!) + \binom{3}{2} \cdot \binom{4}{2} (2!)^2 + 3 \cdot \left(\binom{4}{2} \cdot 2 \right) \cdot (2! \cdot 1! \cdot 1!) = 3 \cdot 1 \cdot 24 + 6 \cdot 4 \cdot 6 + 3 \cdot 6 \cdot 4 + 3 \cdot 12 \cdot 2 = 360$.

The 3 cases are 4 rings on a single finger; $3, 1$ or $2, 2$ rings on two fingers; or $2, 1, 1$ rings on three fingers. In each case, we need to (i) decide which fingers get the numbers of rings, (ii) decide which rings go on each finger, and (iii) order the rings on each finger.

Problem 5.40 Let $\Omega = \{a, b, c\}$ be a sample space, where $\Pr(a) = \frac{1}{2}, \Pr(b) = \frac{1}{3}, \Pr(c) = \frac{1}{6}$. Find the probabilities of all eight possible events.

Solution: $\Pr(\emptyset) = 0, \Pr(\{a\}) = \frac{1}{2}, \Pr(\{b\}) = \frac{1}{3}, \Pr(\{c\}) = \frac{1}{6}, \Pr(\{a, b\}) = \frac{5}{6}$, $\Pr(\{a, c\}) = \frac{2}{3}, \Pr(\{b, c\}) = \frac{1}{2}, \Pr(\{a, b, c\}) = 1$.

Note the 8 possible events are the $2^3 = 8$ subsets of Ω. As an example of calculating probabilities, $\Pr(\{a, b\}) = \Pr(\{a\}) + \Pr(\{b\}) = \frac{1}{2} + \frac{1}{3} = \frac{5}{6}$. The other cases are handled similarly.

Problem 5.41 A die is loaded in such a way that the probability of each face turning up is proportional to the number of dots on that face. (For example, the probability of getting a six is twice that of getting a three.) What is the probability of getting an odd number in one throw?

Solution: $3/7$.

Let $x = \Pr(1)$. Then $\Pr(k) = kx$ for $k = 2, 3, 4, 5, 6$. Hence $\Pr(1) + \Pr(2) + \cdots + \Pr(6) = x + 2x + \cdots + 6x = 21x = 1$ so $x = 1/21$. Hence $\Pr(\{1, 3, 5\}) = (1 + 3 + 5)/21 = 3/7$.

Problem 5.42 Let A and B be events where $\Pr(A) = 1/2, \Pr(\overline{B}) = 1/3$, and $\Pr(A \cap B) = 1/4$. Find the value of $\Pr(A \cup B)$.

Solution: $11/12$.

$\Pr(\overline{B}) = 1/3$ so $\Pr(B) = 2/3$. Then $\Pr(A \cup B) = \Pr(A) + \Pr(B) - \Pr(A \cap B) = 1/2 + 2/3 - 1/4 = 11/12$.

Problem 5.43 Two fair dice are rolled once, and the results are added. What is the probability that the sum is a prime number?

Solution: $5/12$.

There are $6^2 = 36$ total outcomes. Let X denote the sum of the two dice. We have $\Pr(X = 2) = 1/36, \Pr(X = 3) = 2/36, \Pr(X = 5) = 4/36, \Pr(X = 7) = 6/36, \Pr(X = 11) = 2/36$ (for example, we have $3 = 1 + 2, 2 + 1, 5 = 1 + 4, 2 + 3, 3 + 2, 4 + 1$. Thus the probability of X being a prime number is $(1 + 2 + 4 + 6 + 2)/36 = 5/12$.

Problem 5.44 A reader of Marilyn vos Savant's column wrote in with the following question:

> My dad heard this story on the radio. At Duke University, two students had received A's in chemistry all semester. But on the night before the final exam, they were partying in another state and didn't get back to Duke until it was over. Their excuse to the professor was that they had a flat tire, and they asked if they could take a make-up test. The professor agreed, wrote out a test and sent the two to separate rooms to take it. The first question (on one side of the paper) was worth 5 points, and they answered it easily. Then they flipped the paper over and found the second question, worth 95 points: 'Which tire was it?' What was the probability that both students would say the same thing? My dad and I think it's 1 in 16. Is that right?

Is the reader's answer correct? If yes, explain. If no, give the right answer.

Solution: No. It should be $1/4$.

Remember there was really no flat tire! The probability that they both pick the same *specific* tire (that is the front left tire) is $1/16$. However, there are 4 different tires to choose from, so the probability is $4 \cdot (1/16) = 1/4$. Alternatively, the first student can pick any tire and then the second student has a $1/4$ chance of picking the same tire.

Problem 5.45 Events A and B are independent, events A and C are disjoint, and events B and C are independent. If $\Pr(A) = \dfrac{1}{2}$, $\Pr(B) = \dfrac{1}{4}$, $\Pr(C) = \dfrac{1}{8}$, find $\Pr(A \cup B \cup C)$.

Solution: $\dfrac{23}{32}$.

A and C are disjoint, so $\Pr(A \cap C) = 0$ and also $\Pr(A \cap B \cap C)) = 0$. Since A, B and B, C are independent, we have $\Pr(A \cap B) = \dfrac{1}{2} \cdot \dfrac{1}{4} = \dfrac{1}{8}$ and $\Pr(B \cap C) = \dfrac{1}{4} \cdot \dfrac{1}{8} = \dfrac{1}{32}$. Hence (using PIE), we have

$$\Pr(A \cup B \cup C) = \frac{1}{2} + \frac{1}{4} + \frac{1}{8} - \frac{1}{8} - 0 - \frac{1}{32} + 0 = \frac{23}{32}$$

as our final answer.

Problem 5.46 Two numbers X and Y are chosen at random, and with replacement, from the set $\{1,2,3,4,5,6,7,8,9\}$. Find the probability that $X^2 - Y^2$ is an even number.

Solution: $\frac{41}{81}$.

Note that $X^2 - Y^2$ is even if and only if X, Y are both even or both odd. There are 4^2 and 5^2 ways for these to happen, out of 9^2 total outcomes. Hence the probability is $\frac{4^2 + 5^2}{9^2} = \frac{41}{81}$.

Problem 5.47 Three dice are rolled once. Find the probability of getting at least one six.

Solution: $1 - \left(1 - \frac{1}{6}\right)^3 = \frac{91}{216}$.

Calculate the probability of getting no sixes and subtract from one. Assuming each die is fair, the probability of not getting a six on any roll is $1 - \frac{1}{6}$. Since all the rolls are independent, the probability of getting no sixes is $\left(1 - \frac{1}{6}\right)^3 = \frac{125}{216}$.

Problem 5.48 Let A and B be two events and $\Pr(A) = 0.6$, $\Pr(B) = 0.7$. What are the maximum and minimum possible values of $\Pr(A \cap B)$?

Solution: 0.6 and 0.3.

We always have that $A \cap B \subseteq A$ and $A \cap B \subseteq B$, so the maximum value of $\Pr(A \cap B)$ occurs when $A \cap B = A$ so $\Pr(A \cap B) = 0.6$. It is initially plausible that A and B could be disjoint (so $\Pr(A \cap B)$ would equal 0) but this would imply that $\Pr(A \cup B) = 1.3$. Since the maximum possible probability is 1, $\Pr(A \cap B)$ must be at least $0.6 + 0.7 - 1 = 0.3$.

Problem 5.49 Given ten cards with numbers $1, 2, \ldots, 10$ on them, each with one number. Randomly select 3 of the ten cards. What is the probability that ...

  (a) the smallest number is 5?

Solution: $\binom{5}{2} / \binom{10}{3} = 1/12$.

The order of the cards doesn't matter, so the total outcomes is $\binom{10}{3}$. If the smallest number is 5, one of the cards *must* be 5, and the other two numbers are chosen from $6, 7, \ldots, 10$, so $\binom{5}{2}$ outcomes.

(b) the largest number is 5?

Solution: $\binom{4}{2} \Big/ \binom{10}{3} = 1/20$.

Similar to part (a), except the two numbers (other than 5) are chosen from $1, 2, 3, 4$, so $\binom{4}{2}$ outcomes.

Problem 5.50 There are 17 marbles in a bag. Among the marbles, 10 are white, 4 are black, and 3 are red. Nine marbles are taken out at random. What is the probability that 4 white, 3 black, and 2 red marbles are taken out?

Solution: $\dfrac{\binom{10}{4}\binom{4}{3}\binom{3}{2}}{\binom{17}{9}} = \dfrac{252}{2431}$.

There are 17 total balls, so $\binom{17}{9}$ ways to pick 9 of the balls without order. If we want 4 white, 3 black, and 2 red marbles, we have $\binom{10}{4}\binom{4}{3}\binom{3}{2}$ ways to pick (pick which balls for each color).

Problem 5.51 Given 5 distinct pairs of shoes, and randomly select 4 shoes among them. What is the probability that at least one pair of shoes are chosen?

Solution: $1 - \binom{5}{4} \cdot 2^4 \Big/ \binom{10}{4} = \dfrac{13}{21}$.

Consider the complement, which is getting no pairs of shoes. Thus, all the shoes must come from 5 different pairs (so $\binom{5}{4}$ choices for which of the pairs). For each of these 4 chosen pairs, we can choose either the left or right shoe (2 ways for each of the four pairs, so 2^4 in all). Since there are 10 total shoes, we have $\binom{10}{4}$ total ways of picking 4 shoes.

Problem 5.52 Put the 11 letters of "Probability" on 11 cards, and draw 7 cards successively. Find the probability that the 7 cards, in the order drawn, form the word "ability".

Solution: $\dfrac{1 \cdot 2 \cdot 2 \cdot 1 \cdot 1 \cdot 1 \cdot 1}{{}_{11}P_7} = \dfrac{1}{1980}$.

The total number of ways to draw 7 cards is ${}_{11}P_7$. There are two ways to draw "b", and two ways to draw the first "i". Thus the answer is $\dfrac{1\cdot 2\cdot 2\cdot 1\cdot 1\cdot 1\cdot 1}{_{11}P_7}$.

Problem 5.53 Randomly put 3 distinct balls into 4 distinct cups. What is the probability that ...

(a) the largest number of balls in each cup is 1.

Solution: $\dfrac{3}{8}$.

The total number of outcomes is $4^3 = 64$. There are 4 ways for 3 of the four cups to have $= 1$ ball (choose the empty cup). Then there are 3! ways to choose which ball goes in the chosen cups. Hence the probability is $\dfrac{24}{64} = \dfrac{3}{8}$.

(b) the largest number of balls in each cup is 2.

Solution: $9/16$.

If one cup has 2 balls in it, the only case is there are 2 balls in one cup (4 choices for which cup) and 1 ball in a third cup (3 choices). We then decide which two balls go in the first cup (so $\binom{3}{2} = 3$ ways) and the last ball goes in the other cup. Hence the probability is $\dfrac{4\cdot 3\cdot 3}{64} = \dfrac{36}{64} = \dfrac{9}{16}$.

(c) the largest number of balls in each cup is 3.

Solution: $1/16$.

This means all the balls are in a single cup (so 4 choices which cup). The probability is thus $\dfrac{4}{64} = \dfrac{1}{16}$.

Problem 5.54 Let A and B be events, and $\Pr(\overline{A}) = 0.3, \Pr(B) = 0.4, \Pr(A \cap \overline{B}) = 0.5$, find $\Pr(B|(A \cup \overline{B}))$.

Solution: 0.25.

$\Pr(A) = 1 - .3 = .7$, so $\Pr(A \cap B) = \Pr(A) - \Pr(A \cap \overline{B}) = .7 - .5 = .2$. We also have $\Pr(\overline{A} \cap \overline{B}) = 1 - \Pr(B) - \Pr(A \cap \overline{B}) = 1 - .4 - .5 = .1$. Hence, $\Pr(B|A \cup \overline{B}) = \Pr(B \cap (A \cup \overline{B}))/\Pr(A \cup \overline{B})$. Since $\Pr(B \cap (A \cup \overline{B})) = \Pr(A \cap B) = .2$ and $\Pr(A \cup \overline{B}) = \Pr(A) + \Pr(\overline{A} \cap \overline{B}) = .7 + .1 = .8$ the probability we want is $\dfrac{.2}{.8} = \dfrac{1}{4} = 0.25$.

Problem 5.55 Two dice are rolled once. Given that the sum of the two values is 7, determine the probability that one of the dice shows one.

Solution: $1/3$.

There are 6 outcomes $(1,6),(2,5),\ldots,(6,1)$ where the sum of the dice is 7. Only 2 outcomes contain a one, so the probability is $2/6 = 1/3$.

Problem 5.56 Based on statistics, certain family (Dad, Mom, and one Child) has the following probabilities for catching some contagious disease: the probability that the Child is sick is 0.6; the probability that Mom is sick given that Child is sick is 0.5; the probability that Dad gets sick given that Mom and Child are both sick is 0.4. Determine the probability that Mom and Child are both sick but Dad is not.
Solution: 0.18.
Let C, M, D stand for the events that the child, mom, dad are (respectively) sick. The information given is thus $\Pr(C) = 0.6, \Pr(M|C) = 0.5, \Pr(D|(M \cap C)) = 0.4$. So $\Pr(M \cap C) = \Pr(M|C)\Pr(C) = 0.3$, and $\Pr(M \cap C \cap D) = \Pr(D|(M \cap C))\Pr(M \cap C) = 0.12$. Thus $\Pr(M \cap C \cap \overline{D}) = 0.3 - 0.12 = 0.18$.

Problem 5.57 Ten coins are in a bag, two of which are counterfeit. Choose one coin, without putting back, then choose another. What is the probability of each of the following events:

(a) Both are authentic.
   Solution: 28/45.
   We pick the first coin and then the second coin. Hence, there are $10 \cdot 9 = 90$ total outcomes. Similarly, $8 \cdot 7 = 56$ of these outcomes have both coins authentic. Hence the probability is $56/90 = 28/45$.

(b) Both are counterfeit.
   Solution: 1/45.
   There are still 90 total outcomes. Now only $2 \cdot 1 = 2$ of them have two counterfeit coins. Hence the probability is $2/90 = 1/45$.

(c) One authentic, and one counterfeit.
   Solution: 16/45.
   We can either have one authentic and then one counterfeit or vice-versa, so there are $8 \cdot 2 + 2 \cdot 8 = 32$ outcomes we want. Hence the probability is $32/90 = 16/45$.

(d) The second coin is counterfeit.
   Solution: 1/5.
   Consider two cases based on whether the first coin is authentic or not. Hence the number of outcomes with the second coin counterfeit are $8 \cdot 2 + 2 \cdot 1 = 18$. Thus the probability is $18/90 = 1/5$. Note that this is the same as the probability the first coin is counterfeit!

Problem 5.58 You want to call a friend, but forgot the last digit of the phone number. You decide to try dialing the last digit randomly.

(a) What is the probability that you succeed in at most 3 tries?

Solution: $3/10$.

Consider the complement (which is you fail to get the right digit all 3 tries). Assuming you do not try the same last digit more than once, there are $10 \cdot 9 \cdot 8$ total ways you can guess the last number. There are $9 \cdot 8 \cdot 7$ ways to fail (similar to the total except also removing the correct digit). Hence the probability you fail is $\dfrac{9 \cdot 8 \cdot 7}{10 \cdot 9 \cdot 8} = \dfrac{7}{10}$. Thus the probability you succeed is $\dfrac{3}{10}$.

(b) Suppose you know that the last digit is an odd number. What is the probability that you succeed in at most 3 tries?

Solution: $3/5$.

Proceed similarly to part (a) except we now have 5 (instead of all 10) choices for the last number. Hence the probability you fail is $\dfrac{4 \cdot 3 \cdot 2}{5 \cdot 4 \cdot 3} = \dfrac{2}{5}$ so the probability you succeed is $\dfrac{3}{5}$.

Problem 5.59 Two bags are given: the first bag contains n white balls and m red balls; the second bag contains N white balls and M red balls. Take one ball from the first bag and put into the second bag, then take one ball from the second bag. What is the probability that this ball is white?

Solution: $\dfrac{N+1}{M+N+1} \cdot \dfrac{n}{n+m} + \dfrac{N}{M+N+1} \cdot \dfrac{m}{n+m}$.

Let B_1 be the event that the ball from the first bag is white, and B_2 be the event that the ball from the first bag is red. Let A be the event that the ball from the second bag is white. Then by the Total Probability Formula,

$$
\begin{aligned}
\Pr(A) &= \Pr(A|B_1)\Pr(B_1) + \Pr(A|B_2)\Pr(B_2) \\
&= \frac{N+1}{M+N+1} \cdot \frac{n}{n+m} + \frac{N}{M+N+1} \cdot \frac{m}{n+m}.
\end{aligned}
$$

Problem 5.60 Two boxes are given, where the first box contains 5 red balls and 4 white balls, and the second box contains 4 red balls and 5 white balls. Take 2 balls from the first box to put into the second box, then take one ball from the second box. What is the probability that this ball is white?

Solution: $53/99$.

Let B_k for $k = 0,1,2$ be the event that k white balls are chosen from the first box and placed in the second box (so $2-k$ red balls are chosen). (This means there are then 11 balls in the second box, and $5,6,7$ of them are white.) There is no order associated with picking the first two balls, so we have $\Pr(B_k) = \dfrac{\binom{5}{2-k}\binom{4}{k}}{\binom{9}{2}}$. Hence

$\Pr(B_0) = \dfrac{10}{36}, \Pr(B_1) = \dfrac{20}{36}, \Pr(B_2) = \dfrac{6}{36}$. Let A be the event that a white ball is chosen from the second box. Using the Total Probability Formula,

$$\Pr(A) = \Pr(A|B_0)\Pr(B_0) + \Pr(A|B_1)\Pr(B_1) + \Pr(A|B_2)\Pr(B_2)$$
$$= \frac{5+0}{9+2}\cdot\frac{5}{18} + \frac{5+1}{9+2}\cdot\frac{5}{9} + \frac{5+2}{9+2}\cdot\frac{1}{6} = \frac{53}{99}$$

as our final probability.

Problem 5.61 The trade mark of a certain product is "MAXAM", which is mounted on the wall. However, two of the letters fell off and janitor picked up the two letters and randomly put them back on. What is the probability that the sign still says "MAXAM"?
Solution: 3/5.
Note there are $\binom{5}{2} = 10$ collections of two letters that could fall off. Only two of these (AA or MM) result in the same letter falling off. Thus, if B_1 is the event that the same letter fell off twice, then $\Pr(B_1) = \frac{2}{10} = \frac{1}{5}$. Hence if B_2 is that two different letters fell off, $\Pr(B_2) = \frac{4}{5}$. If A is the event that the letters randomly put back on result in the original word "MAXAM", then $\Pr(A|B_1) = 1$ and $\Pr(A|B_2) = \frac{1}{2}$ (the letters are either correct or reversed). Hence using total probability,

$$\Pr(A) = 1\cdot\frac{1}{5} + \frac{1}{2}\cdot\frac{4}{5} = \frac{3}{5}$$

as needed.

Problem 5.62 Suppose that 5% of men are color blind, and 0.25% of women are color blind. From a crowd of equal number of men and women, select one person randomly, and this person happens to be color blind. What is the probability that this is a man?
Solution: 20/21.
Let M, F be the events you pick a man/female and C be the event that the person is color blind. We have $\Pr(M) = \Pr(F) = 0.5$ and $\Pr(C|M) = 0.05, \Pr(C|W) = 0.0025$. Using Bayes' Formula we have

$$\Pr(M|C) = \frac{\Pr(C|M)\Pr(M)}{\Pr(C|M)\Pr(M) + \Pr(C|F)\Pr(F)} = \frac{(0.05)(0.5)}{(0.05)(0.5) + (0.0025)(0.5)} = \frac{20}{21}.$$

as the probability the person who is color blind is a man.

Problem 5.63 A student takes the AMC 12A and 12B. If he passes at least one of these tests, he will qualify for AIME. Assume the probability that he passes the AMC 12A is

p. If he passes the AMC 12A, then the probability that he also passes the AMC 12B is also p. If he fails the AMC 12A, because of the stress, his chance of passing the AMC 12B is only $p/2$.

(a) What is the probability that he is qualified for AIME?

Solution: $\frac{3}{2}p - \frac{1}{2}p^2$.

Let A be the event he passes the AMC 12A and B the event he passes the 12B. The probability he qualified for the AIME is $P(A \cup B) = 1 - P(\overline{A} \cap \overline{B}) = 1 - P(\overline{B}|\overline{A})P(\overline{A})$ which is $1 - (1-p)(1-p/2) = 3p/2 - p^2/2$.

(b) Given that he has passed the AMC 12B. What is the probability that he passed the AMC 12A?

Solution: $\frac{2p}{p+1}$.

Use A, B as above. Using Bayes Formula we have $P(A|B) = \frac{p \cdot p}{p \cdot p + (1-p) \cdot (p/2)} = \frac{2p}{1+p}$.

Problem 5.64 (Monty Hall Problem, based on Marilyn vos Savant's column) Suppose you're on a game show, and you're given the choice of three doors: Behind one door is a car; behind the others, goats. You want to choose the car. You pick a door, say No. 1 (but the door is not opened), and the host, who knows what's behind the doors, opens another door, say No. 3, which has a goat. He then says to you, "Do you want to pick door No. 2?" Is it to your advantage to switch your choice?

Solution: It is; the probability of getting the car is increased from $1/3$ to $2/3$ if you switch.

Note the key idea in the problem is that the host *will not* reveal door one if it contains a car. Let A, B be the event that the car is behind door one or two respectively. Let C be the event that the host reveals door C. We then have that $\Pr(C|A) = 1/2$ while $\Pr(C|B) = 1$ (if No. 1 has the car, it doesn't matter which of the remaining doors the host reveals but if No. 2 has the car, the host must reveal No. 3). As the host will not reveal door No. 3 if it contains the car, we have $\Pr(C) = 1 \cdot \frac{1}{3} + \frac{1}{2} \cdot \frac{1}{3} = \frac{1}{2}$.

Hence we have $\Pr(A|C) = \frac{\Pr(C|A)\Pr(A))}{\Pr(C)} = \frac{(1/2)(1/3)}{(1/2)} = \frac{1}{3}$ and similarly $\Pr(B|C) = \frac{\Pr(C|B)\Pr(B))}{\Pr(C)} = \frac{(1)(1/3)}{(1/2)} = \frac{2}{3}$. so it is in your best interest to switch.

Problem 5.65 Given a line segment of length 1, find two points X and Y on it to split the segment into three pieces. What is the probability that the middle piece is shorter than 1/3?

Solution: 5/9.

Consider all the outcomes as points (x, y) in the box $0 \leq x, y \leq 1$ in the xy-plane. By symmetry we may assume $X < Y$ so we can restrict to points in the box where $y > x$ (i.e. the upper triangle). This region has area $1/2$. If $X < Y$, then the middle piece has length $Y - X$, so we want points (x, y) where $y < x + 1/3$. The following diagram helps illustrate the region:

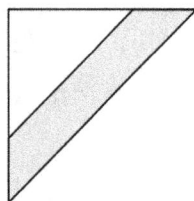

This region has area $\dfrac{1}{2} - \dfrac{1}{2} \cdot \left(\dfrac{2}{3}\right)^2 = \dfrac{5}{18}$, so the probability is $\dfrac{(5/18)}{(1/2)} = \dfrac{5}{9}$.

Problem 5.66 Suppose a bag contains 7 red, 6 green, and 5 yellow balls. Suppose you pick 5 balls at once (so in no particular order).

(a) What is the probability you get 3 red, 1 green, and 1 yellow ball?

Solution: $\dbinom{7}{3} \dbinom{6}{1} \dbinom{5}{1} / \dbinom{18}{5}$.

Since we are picking with no order (and without replacement) there are $\dbinom{18}{5}$ total outcomes. For what we want, we pick which red, green, and yellow balls are chosen.

(b) What is the probability you get 0 yellow balls?

Solution: $\dbinom{13}{5} / \dbinom{18}{5}$.

Proceed similarly to part (a), but thing of the bag as having 5 yellow balls and 13 non-yellow balls. If we want 0 yellow balls all 5 of the balls must be non-yellow.

(c) What is the probability you get all 5 balls of the same color?

Solution: $\left[\dbinom{7}{5} + \dbinom{6}{5} + \dbinom{5}{5} \right] / \dbinom{18}{5}$.

Again proceed similarly to parts (a) and (b), but note that there are 3 cases (all red, all green, or all yellow) to consider.

(d) What is the probability you do not get all red and green balls?

Solution: $1 - \left[\dbinom{13}{5} / \dbinom{18}{5} \right]$.

We do not want all red and green balls. This is equivalent to saying we do not want 0 yellow balls. Hence the probability is 1 minus the probability of 0 yellow balls, which we calculated in part (b).

(e) What is the probability you get at least 2 red, and at least 1 green and at least 1 yellow ball? Hint: Be careful here!

Solution: $\left[\binom{7}{3}\binom{6}{1}\binom{5}{1} + \binom{7}{2}\binom{6}{2}\binom{5}{1} + \binom{7}{2}\binom{6}{1}\binom{5}{2}\right] / \binom{18}{5}$.

This is similar to the above parts, but we do need to consider 3 different cases: (i) 3 red, 1 green, 1 yellow, (ii) 2 red, 2 green, 1 yellow, or (iii) 2 red, 1 green, 2 yellow.

Problem 5.67 Suppose you flip a coin 8 times. What is the probability you get more heads than tails?

Solution: $\dfrac{93}{256}$.

There are 2 outcomes for each flip, so there are $2^8 = 256$ total outcomes. Of these outcomes, $\binom{8}{4} = 70$ have exactly 4 heads and 4 tails. Thus, $256 - 70 = 186$ outcomes have an unequal amount of heads and tails. By symmetry, half of these will have more heads and the other half will have more tails. Thus, there are $186 \div 2 = 93$ outcomes with more heads than tails.

Problem 5.68 Suppose you are dealt a five card hand from a standard deck of 52 cards (with 13 ranks $2,3,\cdots,10,J,Q,K,A$ and 4 suits: hearts, diamonds, clubs, and spades). Find the probability of:

(a) a flush (all cards of the same suit).

Solution: $\binom{4}{1} \cdot \binom{13}{5} / \binom{52}{5}$.

Assume the cards are dealt without order, so there are $\binom{52}{5}$ total outcomes. We first choose the suit $\binom{4}{1}$ ways. There are then 13 cards of that suit (the 13 different ranks) so we choose 5 of them (again without order, so $\binom{13}{5}$ ways).

(b) a straight (all 5 ranks in a row, so $A,2,3,4,5$ up to $10,J,Q,K,A$).

Solution: $10 \cdot 4^5 / \binom{52}{5}$.

There are still $\binom{52}{5}$ total outcomes. Note the ranks of the cards in a straight are determined by the lowest card (which can be $A,2,3,\ldots,10$), so there are 10 choices for the ranks of the cards for the straight. Each rank can be any of the 4 suits (so 4^5 total).

(c) a full house (3 cards of one rank, 2 cards of another).

Solution: $\binom{13}{1}\binom{12}{1}\binom{4}{3}\binom{4}{2}/\binom{52}{5}$.

There are still $\binom{52}{5}$ total outcomes. We first choose the rank for the three of a kind, then the rank of the pair. There are a total of $\binom{13}{1}\binom{12}{1}$ ways to do so. Then we choose the 3 suits for the three of a kind (chosen out of the 4 suits total, so $\binom{4}{3}$ ways), then the 2 suits for the pair (so $\binom{4}{2}$ ways).

(d) a four of a kind.

Solution: $\binom{13}{1}\binom{12}{1}\binom{4}{4}\binom{4}{1}/\binom{52}{5}$.

There are still $\binom{52}{5}$ total outcomes. We still have 5 total cards, so 1 card is a single. We first pick the rank of the four of a kind, then the final rank (so $\binom{13}{1}\binom{12}{1}$ ways). We then choose the suits, first for the four of a kind and then for the single (so $\binom{4}{4}\binom{4}{1}$ ways together).

(e) exactly two pairs. (That is, not four of a kind or a full house.)

Solution: $\binom{13}{2}\binom{11}{1}\binom{4}{2}^2\binom{4}{1}/\binom{52}{5}$.

There are still $\binom{52}{5}$ total outcomes. Pick the ranks of the pairs at once, $\binom{13}{2}$ ways, because "a pair of J and a pair of Q" is the same as "a pair of Q and a pair of J". Then pick the final rank from the remaining 11 ranks, $\binom{11}{1}$ ways. Then choose the suits for the two pairs and the single, $\binom{4}{2}^2\binom{4}{1}$.

(f) exactly one pair. (That is, not two pair or three of a kind, etc.)

Solution: $\binom{13}{1}\binom{12}{3}\binom{4}{2}\binom{4}{1}^3/\binom{52}{5}$.

There are still $\binom{52}{5}$ total outcomes. Pick the rank of the pair, $\binom{13}{1}$ ways. Then pick the ranks of the singles at once, $\binom{12}{3}$ ways. Then choose the suits for the pair and the three single, $\binom{4}{2}\binom{4}{1}^3$.

Problem 5.69 Suppose you randomly pack a cooler (in no particular order) of 40 sodas (chosen from Coke, Sprite, and Fanta). Find the probability that:

(a) there are 20 Cokes and 20 Sprites.

Solution: $1/\binom{40+3-1}{40}$.

There is only one outcome with 20 Cokes and 20 Sprites. We use stars and bars to calculate the total outcomes.

(b) there are equal amounts of all sodas.

Solution: 0.

It is impossible to have an equal amount of all sodas as 40 is not a multiple of 3.

(c) there are at least 10 Cokes.

Solution: $\binom{30+3-1}{30}/\binom{40+3-1}{40}$.

We know at least 10 must be Coke. This leaves 30 sodas which can be chosen among any of the 3 types.

(d) there are at least 5 of each soda.

Solution: $\binom{25+3-1}{25}/\binom{40+3-1}{40}$.

After there are 5 of each type, there are 25 remaining sodas which can be chosen among any of the 3 types.

Problem 5.70 Suppose Jack and Jill both randomly come to school between 10 AM and 2 PM.

(a) What is the probability Jack comes to school before 11 AM?

Solution: $\frac{1}{4}$.

Consider the outcomes on a number line. There are 4 total hours, and 1 of those hours is before 11 AM, so the probability is $\frac{1}{4}$.

(b) What is the probability Jack and Jill both come to school before 11 AM?

Solution: $\frac{1}{16}$.

Consider a square in the xy-plane, where the x-axis corresponds to when Jack arrives and the y-axis corresponds to when Jill arrives. This gives the picture below

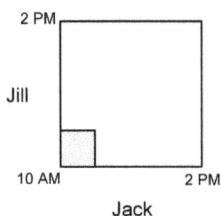

where the shaded region is Jack and Jill both coming to school before 11 AM. This region is 1/16th of the entire square.

(c) What is the probability Jack comes to school before Jill?

Solution: $\frac{1}{2}$.

Use a method similar to part (b). Note that Jack arriving before Jill corresponds to the region where $x < y$. We have the picture

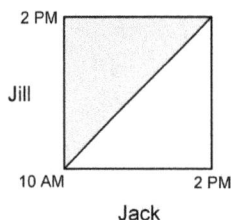

where the shaded region is Jack coming to school before Jill. This region is 1/2 of the entire square.

(d) What is the probability Jack and Jill come to school within 30 minutes of each other?

Solution: $\frac{15}{64}$.

Use a method similar to part (b). Note that Jack and Jill coming within 30 minutes of each other is the same as: (i)Jack coming before Jill or up to 30 minutes after Jill, corresponding to the region $x < y + 1/2$ AND (ii) Jill coming before Jack or up to 30 minutes after Jill, corresponding to the region $x > y - 1/2$. Hence, we have the picture below

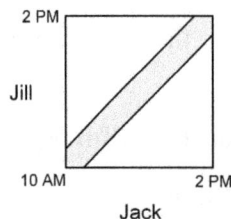

where now the shaded region is Jack and Jill arriving within 30 minutes of each other. This region is 15/64 of the entire square.

Problem 5.71 Suppose Jackie is at a track meet. She is an erratic long jumper: each of her jumps are randomly between 2 and 8 meters.

(a) What is the probability a jump is more than 7 meters?

Solution: $\frac{1}{6}$.

There are 6 meters in total (length), and 1 of these meters is longer than 7 meters.

(b) What is the probability the jump is a integer length when measured in centimeters?
Solution: 0.

All the integer lengths when measured in centimeters have 0 length.

(c) If Jackie has two tries to jump more than 5 meters, what is the probability she succeeds (at least once)?

Solution: $\frac{3}{4}$.

Jackie has a $3/6 = 1/2$ chance of jumping more than 5 meters (6 meters in total and 3 of these meters correspond to a jump longer than 5 meters). Thus, Jackie has a $(1/2)^2 = 1/4$ chance of failing both times. This is the complement of what we want, hence she has a $3/4$ chance of succeeding at least once.

Problem 5.72 There are 5 balls in a bag, numbered 1,2,3,4,5. Take 3 balls from the bag, and let X represent the largest number on the 3 balls. What is the distribution of X?
Solution: $\Pr(X = 3) = 1/10, \Pr(X = 4) = 3/10, \Pr(X = 5) = 3/5$.

Note there are $\binom{5}{3} = 10$ total outcomes. Since we are picking 3 balls, the largest is either numbered 3, 4, or 5. If it is numbered 3, then we must pick balls 1,2,3 (only one way), hence $\Pr(X = 3) = 1/10$. If the largest is numbered 4, one ball must be 4 and the remaining are picked from balls 1,2,3 (so $\binom{3}{2} = 3$ ways), hence $\Pr(X = 4) = 3/10$. Since $X = 5$ is the only other outcomes, $\Pr(X = 5) = 1 - 1/10 - 3/10 = 6/10 = 3/5$, finishing the distribution of X.

Problem 5.73 Roll a die twice. Find the distribution of the sum of the two values.
Solution: The sum can be 2,3,4,...,12, and the probability for each value is:
$1/36, 2/36, 3/36, 4/36, 5/36, 6/36, 5/36, 4/36, 3/36, 2/36, 1/36$.
There are $6^2 = 36$ total outcomes. If we think of these outcomes in a 6×6 table, note that all the $45°$ diagonals (lower-left to upper-right) have a constant sum. This helps explain the pattern, as is one way for a sum of 2 (outcome of $(1,1)$), two ways for 3 (outcomes of $(2,1),(1,2)$), etc. up to 6 ways for a sum of 7 (outcomes $(1,6),(2,5),...,(6,1)$), and then back down to one way for a sum of 12 (outcomes of $(6,6)$).

Problem 5.74 There are 15 balls in a bag, two of which are red and the remaining are white. Take 3 balls out of the bag. Let X be the number of red balls taken. Find the

distribution of X.
Solution: $X = 0, 1, 2$, and the probabilities are $22/35, 12/35, 1/35$.
There are $\binom{15}{3} = 455$ total outcomes. Since we have 2 red balls, $k = 0, 1, 2$ of them could be red. There are $\binom{2}{k}\binom{11}{3-k}$ ways of choosing which k red balls and thus $3 - k$ white balls there are out of the 2 red and 13 white in total. This gives $286, 156, 13$ outcomes for each case. Dividing each by 455 gives the probabilities for (respectively) $0, 1, 2$ red balls.

Problem 5.75 Keith gets off work everyday at 5pm. He can choose to take the bus or subway to get home. From his experience, the probabilities of getting home at certain time range taking either subway or bus is in the following table:

Arrival time	5:35-5:39	5:40-5:44	5:45-5:49	5:50-5:54	5:55 & later
Subway	0.10	0.25	0.45	0.15	0.05
Bus	0.30	0.35	0.20	0.10	0.05

Today, he decided to flip a coin to choose between bus and subway, and then he arrived home at 5:47. What is the probability that he took the subway?
Solution: $9/13$..
Let A be the event Keith arrives home in the interval 5:45-5:49, and S, B the events he took the subway or bus. Using Bayes' Theorem we have

$$\Pr(S|A) = \frac{\Pr(A|S)\Pr(S)}{\Pr(A|S)\Pr(S) + \Pr(A|B)\Pr(B)} = \frac{0.45 \cdot 0.5}{0.45 \cdot 0.5 + 0.2 \cdot 0.5} = \frac{0.45}{0.65} = \frac{9}{13}$$

for the probability he took the subway given his arrival time.

Problem 5.76 You are playing darts. Suppose the probability that you hit the target is 45%. Let X be the number of throws by the time you first hit the target. Find (1) the distribution of X, and (2) the probability that X is even.
Solution: (1) $\Pr(X = k) = 0.45 \times (0.55)^{k-1}, k = 1, 2, \ldots$.

(2) The probability that X is even is $\frac{11}{31}$. Note this is called a Geometric distribution. For (1), if the kth throw is the first time the target is hit, there must be $k - 1$ misses (each with probability 0.55) and then a hit (with probability 0.45). For (2), we use (1) and the fact that the sum of a geometric sequence $1, r, r^2, r^3, \ldots$ is $1/(1-r)$. Hence the probability X is even is (note here $r = 0.55^2$)

$$\sum_{k=1}^{\infty} \Pr(X = 2k) = 0.45 \times (0.55 + 0.55^3 + 0.55^5 + \cdots) = 0.45 \times 0.55 \times \frac{1}{1 - (0.55)^2} = \frac{11}{31}.$$

Problem 5.77 A room has 3 windows of the same shape and size, only one of which is open. A bird flies into the room through the open window. It can only fly out through the open window. It flies around in the room and tries to find the open window. Assume that this bird does not remember which windows have been tried, and has the same probability to try any of the windows.

(a) Let X be the number of times the bird tries to fly through a window and finally gets out. Find the distribution of X.

Solution: $\Pr(X = k) = (1/3) \cdot (2/3)^{k-1}$.

Note this is a Geometric distribution. To fly (successfully) to fly out of the window on the kth try, the bird must fail $k - 1$ times (with a $2/3$ chance of failing for each of these tries) and then succeed (with a $1/3$ chance).

(b) The owner said he has a bird that can memorize the windows it has tried, thus it tries each window at most once. Assume the owner is telling the truth, let Y be the number of tries this smart bird uses to get out the room. Find the distribution of Y.

Solution: $1/3$ for each of $Y = 1, 2, 3$.

It is clear that $\Pr(Y = 1) = 1/3$. For $Y = 2$, note the bird must first fail ($2/3$ chance) and then successfully pick the correct window of the two remaining ($1/2$ chance). Therefore $\Pr(Y = 2) = (2/3) \times (1/2) = (1/3)$. Since $Y = 3$ is the only remaining possibility, $\Pr(Y = 3) = 1 - (1/3) - (1/3) = (1/3)$ as well.

(c) Calculate $\Pr(X < Y)$.

Solution: $8/27$.

For $Y < X$ we have the following possibilities for (X, Y): $(1, 2), (1, 3), (2, 3)$. We add up these three cases, so $\Pr(X < Y) = (1/3) \times (1/3) + (1/3) \times (1/3) + (2/9) \times (1/3) = 8/27$.

(d) Calculate $\Pr(Y < X)$.

Solution: $38/81$.

Note (considering the three cases $(X, Y) = (1, 1), (2, 2), (3, 3)$ we have $\Pr(X = Y) = (1/3) \times (1/3) + (2/9) \times (1/3) + (4/27) \times (1/3) = 19/81$. Hence $\Pr(Y < X) = 1 - \Pr(X = Y) - \Pr(X < Y) = 1 - 19/81 - 8/27 = 43/81$.

Problem 5.78 Adam and Bob played darts. Adam's probability of hitting the target is 0.6, and Bob's is 0.7. They each throw 3 times. Find the probability that ...

(a) they hit the same number of times.

Solution: 0.32076

If they hit the same number of times, the possibilities are each hitting $0, 1, 2, 3$ times. If they have a p chance of hitting the target (so $p = .6$ or $p = .7$), the probability they hit the board k times is $\binom{3}{k} p^k (1-p)^{3-k}$ for $k = 0, 1, 2, 3$ (we need to decide which of the throws hit the target, and then the probability for each through hitting or missing is p or $(1-p)$ (note this is called a binomial distribution). Hence, the probability Adam and Bob hitting the same number of times is

$$\left((0.6)^0(0.4)^3\right)\left((0.7)^0(0.3)^3\right)+\left(3\times(0.6)^1(0.4)^2\right)\left(3\times(0.7)^1(0.3)^2\right)$$
$$+\left(3\times(0.6)^2(0.4)^1\right)\left(3\times(0.7)^2(0.3)^1\right)+\left((0.6)^3(0.4)^0\right)\left((0.7)^3(0.3)^0\right)$$

which after simplifying is 0.32076.

(b) Adam hits more than Bob.
Solution: 0.243

Note as in part (a), we can use the formula $\binom{3}{k}p^k(1-p)^{3-k}$ for the probability of hitting k times (with $p=.6$ or $p=.7$ for Adam or Bob). If Adam hits more than Bob, then Bob can either hit $0,1,2$ times (so Adam his $1-3,2-3,3$ times). Note Adam hitting between $1-3$ times is the complement of Adam hitting 0 times. Hence the probability is:

$$\left(1-(0.6)^0(0.4)^3\right)\left((0.7)^0(0.3)^3\right)$$
$$+\left(3\times(0.6)^2(0.4)^1+(0.6)^3(0.4)^0\right)\left(3\times(0.7)^1(0.3)^2\right)$$
$$+\left((0.6)^3(0.4)^0\right)\left(3\times(0.7)^2(0.3)^1\right)$$

which after simplifying is 0.243.

Problem 5.79 Normally, we assume that a new born child has a 50-50 chance to be a boy or a girl. Suppose, in a certain country, because of the desires for each family to have sons, a policy was made by the government: Any family with a son already is not allowed to have more children. Any family with all girls is allowed to have one more child, and so on. (1) So roughly among the children, are there more boys or more girls? (2) Find the expected number of children in each family.
Solution: (1) Roughly it's the same. (2) 2.
Let X denote the number of children in a family. Then $\Pr(X=k)=(1/2)^k$ as the family stops having children when they have a boy (as in earlier problems this is a geometric distribution). We can use the following trick to calculate the expected value of X, which

is $1 \cdot (1/2) + 2 \cdot (1/2)^2 + 3 \cdot (1/3)^3 + \cdots = \sum_{k=1}^{\infty} k \times (1/2)^k$:

$$\sum_{k=1}^{\infty} k \times (1/2)^k = \sum_{k=0}^{\infty} (k+1) \times (1/2)^{k+1}$$

$$= \sum_{k=0}^{\infty} k \times (1/2)^{k+1} + \sum_{k=0}^{\infty} (1/2)^{k+1}$$

$$= (1/2) \times \sum_{k=1}^{\infty} + \sum_{k=0}^{\infty} (1/2)^{k+1}$$

$$\Rightarrow (1/2) \times \sum_{k=1}^{\infty} k \times (1/2)^k = \sum_{k=0}^{\infty} (1/2)^{k+1} = (1/2) \times \frac{1}{1-(1/2)}$$

$$\Rightarrow \sum_{k=1}^{\infty} k \times (1/2)^k = 2$$

so the expected number of children in each family is 2.

Second Solution: Alternatively, if the family first has a boy, they stop (after 1 child) with probability $1/2$. Otherwise, the family "starts over" trying to get a boy (after 1 child already). Hence we have the equation $EX = 1 \times \dfrac{1}{2} + (1 + EX) \times \dfrac{1}{2}$ and we can solve for $EX = 2$.

Problem 5.80 Each of three friends decides to go to a fitness club at least once in a 5-day work week. Each of them select a day at random. What is the probability that they go in 3 separate days?

Solution: 12/25.

There are $5^3 = 125$ total ways the friends can each pick a day of the week. For them to go on separate days, the first can choose any day (5 choices), the second can choose any of the remaining 4 days, and the third can choose any of the 3 remaining (so $5 \times 4 \times 3 = 60$ total). Hence the probability is $60/125 = 12/25$.

Problem 5.81 Three hunters shot at a rabbit. If the probabilities of hitting the target was 0.6, 0.5, and 0.4 for each of them respectively, what is the probability that the rabbit was hit?

Solution: 0.88 (or 88%).

For the rabbit to be hit, it needs to be hit at least once. The probability the rabbit is not hit (that is, each hunter misses) is $(1 - 0.6) \times (1 - 0.5) \times (1 - 0.4) = 0.12$. Hence the probability the rabbit is hit is $1 - 0.12 = 0.88$.

Problem 5.82 Only one key is needed to open the door, and David has five keys, two of which can open the door. The problem is that he doesn't know which two. David tries the five keys one by one. What is the probability that he opens the door within three tries?

Solution: 0.9 (or 90%).

The total number of ways to try the first three keys is $5 \times 4 \times 3 = 60$. There are $3 \times 2 \times 1 = 6$ ways to choose the wrong three keys in the first three tries. Hence David has a $6/60 = 1/10$ chance of *not* opening the door, so a 0.9 chance of opening it.

Problem 5.83 A fair coin is flipped repeatedly, until the first heads occurs.

  (a) What is the probability exactly two flips are needed? Three flips? Four flips?
      Solution: $1/4, 1/8, 1/16$.

  (b) What is the probability that exactly n flips are needed?
      Solution: $1/2^n$.
      For n flips to be needed, the first $n-1$ flips must all be tails, and then the nth flip must be heads. This is a single outcome ($T \cdots TH$) out of the 2^n total outcomes (H or T for each flip, hence the probability is $1/2^n$.

Problem 5.84 Same situation as the previous question. What is the expected value of number of flips until heads occurs?

Solution: 2.

See the explanation in Problem 5.79.

Problem 5.85 Frank and Ed play 5 games during a contest. If either of them wins 3 games, the contest is over. If for each game, the probability that Frank wins is $\frac{2}{3}$, and the probability that Ed wins is $\frac{1}{3}$. What is the probability that Frank wins the contest 3 : 1?

Solution: $\frac{8}{27}$.

Note for the contest to have taken 4 games, Frank must have one game 4. Then Ed wins one of the first three games (so there are $\binom{3}{1} = 3$ choices for which game he wins). Once we have decided who wins what games, we use the fact that Frank wins a given game with probability 2/3 (so Ed wins with probability 1/3). Thus, the final probability is $3 \times \left(\frac{2}{3}\right)^3 \left(\frac{1}{3}\right)^1 = \frac{8}{27}$.

Problem 5.86 Two points are randomly selected on a circle. What is the probability that the chord connecting the two points is longer than the radius?

Solution: $2/3$.

Let the two points be P, Q and the two chords be $\overline{PQ}$ and $\overline{RS}$. Note that the probability does not depend on where P is chosen, so we only worry about R. Since $PQ = r$, connecting P, Q to the center of the circle gives an equilateral triangle, hence $\overset{\frown}{PQ}$ is a $60°$-arc. Similar reasoning holds for $\overset{\frown}{RS}$ so R must outside arc $\overset{\frown}{PQ}$ ($60°$ total) and R must be outside the $60°$ arc "before" P (else then P is inside $\overset{\frown}{RS}$). Hence the probability is $\dfrac{240}{360} = \dfrac{2}{3}$.

Problem 5.87 There are 5 distinct red balls and 4 distinct white balls in a bag. Ed takes a ball from the bag and then places it back. He does this three times. What is the probability that he gets a red ball twice and a white ball once?

Solution: $100/243$.

There are $\binom{3}{1} = 3$ orders (RRW, RWR, WRR) Ed can get $= 2$ red balls. On a given pick, Ed has a $5/9$ chance of picking a red ball and a $4/9$ chance of picking a white ball. Hence the final probability is $3 \times (5/9)^2 (4/9)^1 = 100/243$ of picking a red ball twice and a white ball once.

Problem 5.88 Randomly put n distinct balls into n boxes. Find the probability of getting exactly one empty box.

Solution: $\binom{n}{2} \cdot n!/n^n$.

Since all the balls and boxes are distinct, there are n^n total outcomes. If exactly one box is empty, then another box has two balls in it. There are n choices for which box is empty, then $n - 1$ choices for which box has two balls. For this box, there are $\binom{n}{2}$ ways to choose which two balls it contains. We are left with $n - 2$ balls and $n - 2$ boxes, so there are $(n-2)!$ ways to arrange the rest of the balls (one in each box). Hence the final probability is (recall $n(n - 1)(n - 2)! = n!$) is $\binom{n}{2} \cdot n!/n^n$ as needed.

Problem 5.89 Randomly put 3 letters into 3 envelopes that are already addressed. Assume that each letter corresponds to only one correct address. Find the probability that at least one letter is in the correct envelope.

Solution: $2/3$.

Suppose the letters are labeled 1,2,3 and the envelopes 1,2,3. There are $3! = 6$ total arrangements of the letters and envelopes (corresponding to all the orderings of 1,2,3). Only 231 and 312 have the all three letters in the wrong spot, so there is a $4/6 = 2/3$ chance that at least one letter is in the correct envelope.

Problem 5.90 Ten people want to go to a basketball game, but there is only one ticket. So the ticket is put in one of 10 envelopes that look exactly the same, and the ten people pick up the envelopes one by one. Is it better to pick up an envelope earlier or later?
<u>Solution</u>: Neither. Everyone has 1/10 probability to get the ticket.

Problem 5.91 A fair coin is flipped repeatedly until two heads in a row appears. What is the probability that exactly 4 flips are needed?
<u>Solution</u>: 1/8.
There are $2^4 = 16$ total ways to flip a coin 4 times. If the fourth flip was the first time 2 heads appeared in a row, the only possibilities for the 4 flips are $HTHH$ or $TTHH$. Hence the probability is $2/16 = 1/8$.

Problem 5.92 Same situation as the previous question. What is the expected value of number of flips until two heads in a row occurs?
<u>Solution</u>: 6.
Note if we get 2 heads in a row (1/4 chance) we are done after 2 flips. Once we get a tail (either T or HT with probability $1/2, 1/4$ respectively), we "start over" after 1 or 2 flips. Hence we have the equation

$$EX = 2 \times \frac{1}{4} + (1 + EX) \times \frac{1}{2} + (2 + EX) \times \frac{1}{4}.$$

We can then solve to get $EX = 6$.